AC/DC Electricity and Electronics Made Easy

2nd Edition

Victor F. Veley

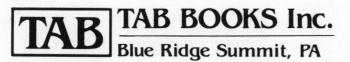

TAB BOOKS Inc.
Blue Ridge Summit, PA

SECOND EDITION
FIRST PRINTING

Library of Congress Cataloging in Publication Data

Veley, Victor F. C.
AC/DC electricity and electronics made easy / by Victor F. Veley.
—2nd ed.
p.   cm.
ISBN 0-8306-9285-1   ISBN 0-8306-3285-9 (pbk.)
1. Electricity. 2. Electronics. I. Title.
QC522.V44   1990
621.3—dc20   89-20471
                                   CIP

TAB BOOKS Inc. offers software for
sale. For information and a catalog,
please contact TAB Software Department,
Blue Ridge Summit, PA 17294-0850.

Questions regarding the content of this book
should be addressed to:

Reader Inquiry Branch
TAB BOOKS Inc.
Blue Ridge Summit, PA 17294-0214

Acquisitions Editor: Roland S. Phelps
Technical Editor: Steve Burwen
Production: Katherine Brown
Cover Design: Lori E. Schlosser

Contents

Introduction

This second edition of *AC/DC Electricity and Electronics Made Easy* has been written to expand the breadth of the material presented. To achieve this objective, five new chapters have been added:

1. Chapter 11. "Decibels and Nepers." This is an expansion of the previous Appendix A.
2. Chapter 14. "Filters."
3. Chapter 16. "AC Circuit Analysis."
4. Chapter 18. "Motors (DC and AC)."
5. Chapter 19. "Nonsinusoidal Waveforms." A treatment of this subject is rarely found in introductory texts.

The aims of the second edition in no way differ from those of the first edition. Technicians and students of electronics technology are often required to substitute numerical values in an expression to derive a particular answer. There are only two problems. First of all, it is necessary to remember the formula correctly, and this might not be too easy, especially if there are various powers of ten and constants involved. Second, you must recognize the units in which the answer is obtained and then convert, if required, to practical values. This is certainly true if you are concerned with electromagnetic units (emu) or electrostatic units (esu).

How much simpler it would be if there were only one system of electrical units! In 1901, Professor Giorgi of Italy proposed a system based on the meter, kilogram, and second (mks) for length, mass, and time with the choice of a fourth fundamental unit being the ampere. Since 1950, more countries have officially recognized this system, and there is little doubt that, in the future, it will be adopted universally. The advantages lie in the simpler formulas, which always employ the practical units, and the elimination of the confusion that existed between the previous systems. This was achieved at the expense of introducing certain additional units and assigning new values to the permeability and permittivity of free space.

I had taught electronics for many years using the old systems and was agreeably surprised to find how much simpler the mks system was for the instructor to use and for the student to understand. This book is therefore the outcome of my experience in teaching to a variety of classes the subjects of direct and alternating current.

The purpose of this book is twofold. First, it should prove to be a valuable reference for the technician in the field. Each chapter is organized into a number of sections. In each section, one or more important concepts are introduced with the relevant equations. The concepts and the equations then are used in a number of comprehensive examples, which are accompanied by detailed solutions. Finally, at the end of each chapter is a summary of the more important equations and, where necessary, a table to illustrate the differences between the modern SI units and the older cgs systems. You are therefore given every opportunity to familiarize yourself with the modern system of units.

The second purpose is to assist community college students in the first and second semesters of electronics technology. The survey of SI units in this reference manual will be a valuable supplement to any standard textbook the student is using. In addition, the examples chosen are unusually comprehensive and should provide the student with a better understanding of the principles involved. In virtually every example, only standard practical values are used, and many problems are solved by a variety of methods. This method serves as an accurate check on the answers obtained and as a comparison among various analytical techniques. Furthermore, many circuits can be constructed in the laboratory so that calculated results can be compared with experimental readings.

I wish to acknowledge a great debt to my wife, Joyce, for typing the manuscript and for making this second edition possible.

1
Mechanical Units

IN THIS CHAPTER you will learn about:

☐ The various mechanical S.I. units and how they are interrelated.
☐ The various prefixes and how they are used in electronics.

A system of units represents the language in which a scientific subject is discussed. All such systems are built on certain fundamental units, which are taken as standards.

FUNDAMENTAL UNITS OF THE MKS SYSTEM

In the *mks system,* the unit of mass is the kilogram (1000 grams), which is defined by an artificial standard held by an international bureau in France. The unit of length is the *meter,* which is derived from an atomic standard, while the second is 1/86,400 of the mean solar day.

UNIT OF FORCE

When a force is applied to a mass, the mass accelerates (resulting in a change in velocity). In the mks system, acceleration is measured in meters per second per second. For example, if a mass starts from rest (zero velocity) with a constant acceleration of 2 meters per second per second, its velocity after 1 second is 2 meters per second. After 2 seconds, it is 4 meters per second, and so on. There are many units of force in existence, but in the mks system, a new one is introduced. This is the *newton* (named after Issac Newton, 1642–1727), which is the force required to give a mass of 1 kilogram an acceleration of 1 meter per second per second in the direction of the force. F (force in newtons) = M (mass in kilograms) × A (acceleration in meters per second per second).

The force of gravity gives a mass an acceleration of approximately 9.81 meters per second per second. This is not constant but varies slightly over different positions on the earth's surface. The force exerted on a mass of 1 kilogram due to gravity is 9.81 newtons; this force sometimes is referred to as a 1 kilogram weight. On the moon, the gravitational force on 1 kilogram would be much less than 9.81 newtons.

In the old centimeter-gram-second system, the unit of force was the *dyne,* which gave 1 gram an acceleration of 1 centimeter per second per second. Because 1 kilogram = 1000 gms and 1 meter = 100 centimeters, 1 newton = 100,000 dynes. For most purposes, the dyne is far too small, and therefore the newton has a practical advantage.

UNIT OF ENERGY OR WORK

When a force (newtons) is exerted through a distance (meters) in the direction of the force, work is done or energy is consumed (energy is the capacity for doing work). In the mks system the unit of work is the *joule* (named after James Joule, 1818–1899), which is equivalent to 1 meter-newton.

Work done = F (force in newtons) × D (distance in meters) = F × D (joules or meter-newtons, *not* newtons-meters.) (See unit of torque.)

Because the various forms of energy are interchangeable, the joule also will be the mks unit of electrical energy. In the cgs system, the *erg* is the work done when a force of 1 *dyne* acts through 1 centimeter in the direction of the force. 1 joule = 1 meter-newton = 100 × 100,000 = 10,000,000 ergs.

The unit of heat energy is the *calorie,* which is the amount of heat required to raise the temperature of 1 gram of water by 1° centigrade. It has been found experimentally that 1 calorie = 4.18 joules.

UNIT OF TORQUE

Work is the product of force and distance, but to obtain *torque,* it is also necessary to multiply these quantities together. However, for work the force and distance lie in the same direction, whereas for torque the force and the distance are at right angles to each other. In Fig. 1-1, 0 is a pivotal point and a force F is acting at a point P, distance D from O. The torque of F about O = F × D newton-meters (*not* joules).

In cgs units, torque is measured in dyne-centimeters.

POWER

Power is the *rate* at which work is done or energy is produced. It is sometimes difficult to distinguish between power and work. For example, a child might use a complex system of pulleys to lift a heavy weight against gravity, while a strong man can swiftly lift the weight through the same

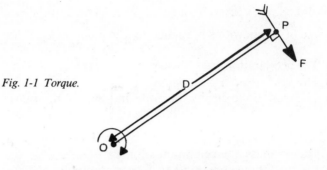

Fig. 1-1 Torque.

distance. Neglecting losses in the pulley system, the man and the child perform the same total work, but because the man took a shorter time, his power is greater. In the mks system, the unit of power is the *watt* (named after James Watt, 1736–1819). It is equal to 1 joule of energy being produced or released per second. A 60-watt electric light bulb dissipates 60 joules of energy every second (mostly in the form of heat but a small amount as light). Watts = joules per second or joules = watts × seconds.

The unit on your electric bill is a measure of the work done. The joule is too small for this purpose, and the unit chosen is the kilowatt-hour (kWh) = 1000 watts × 60 minutes × 60 seconds = 3,600,000 joules.

Another mechanical unit of power is the horsepower, which is equal to 746 watts.

PREFIXES

A number of prefixes are used in electronics to create multiples or submultiples of basic units. The most important are shown in Table 1-1.

Table 1-1.

	Abbreviation	Factor
Tera-	T	× 1,000,000,000,000 or 10^{12}
Giga-	G	× 1,000,000,000 or 10^9
Mega-	M	× 1,000,000 or 10^6
Kilo-	k	× 1,000 or 10^3
Milli-	m	÷ 1,000 or × 10^{-3}
Micro-	μ	÷ 1,000,000 or × 10^{-6}
Nano-	n	÷ 1,000,000,000 or × 10^{-9}
Micro-Micro	μμ or	÷ 1,000,000,000,000 or × 10^{-12}
or Pico-	P	

Example 1-1

A force of 200 newtons is applied to a mass of 50 kg. Find:

(a) The acceleration
(b) The velocity after 5 seconds
(c) The work done during this time and the average power

Solution

(a) force = mass × acceleration
      200 newtons = 50 kg × acceleration

$$\text{acceleration} = \frac{200 \text{ newtons}}{50 \text{ kg}} = \frac{4 \text{ meters per second}}{\text{per second}}$$

(b) velocity after 5 seconds = acceleration × time
                    = 4 meters per second per second × 5 seconds
                    = 20 meters per second

(c) average velocity $= \dfrac{20}{2} = 10$ meters per second

  distance covered = average velocity × time
                = 10 meters per second × 5 seconds
                = 50 meters
  work done = 200 newtons × 50 meters
                = 10,000 joules

$$\text{average power} = \frac{10,000 \text{ joules}}{5 \text{ seconds}}$$

                = 2000 watts
                = 2 kW

Example 1-2

A metal block whose mass is 180 grams is given an acceleration of 25 cm/sec^2. What is the accelerating force in newtons? What is the kinetic energy in joules when the block's velocity is 450 cm/sec?

Solution

$$\text{force} = \text{mass (kg)} \times \text{acceleration (m/sec}^2)$$

$$= \frac{180}{1000} \times \frac{25}{100}$$

$$= 0.045 \text{ newton}$$

$$\text{kinetic energy} = \frac{1}{2} \times \text{mass (kg)} \times \text{velocity (m/sec)}^2$$

$$= \frac{1}{2} \times \frac{180}{1000} \times \left(\frac{450}{100}\right)^2$$

$$= 1.82 \text{ joules}$$

Example 1-3

A mass of 1600 kg is lifted vertically with a velocity of 90 meters per minute. Calculate

(a) The power required in kilowatts
(b) The kinetic energy of the load in joules

Solution

$$\text{force required} = 1600 \times 9.81 \text{ newtons}$$

$$\text{velocity} = \frac{90}{60} = 1.5 \text{ meters per second}$$

$$\text{power} = 1600 \times 9.81 \times 1.5 \text{ W} = \frac{1600 \times 9.81 \times 1.5}{1000} \text{ kW}$$

$$= 23.5 \text{ kW}$$

$$\text{kinetic energy} = \frac{1}{2} \times 1600 \times (1.5)^2$$

$$= 1800 \text{ joules}$$

Example 1-4

A motor develops 60 horse-power with a speed of 800 rpm. Calculate its torque in newton-meters.

Solution

$$60 \text{ hp} = 60 \times 746 \text{ W} = 44760 \text{ W}$$

$$\text{angular velocity} = 2\pi \times \frac{800}{60} = 83.776 \text{ radians per second}$$

$$\text{torque} = \frac{44760}{83.776} = 534 \text{ newton-meters}$$

Example 1-5

A mass of 175 kg is pulled along a horizontal surface whose coefficient of friction is 0.32. If the mass covers a distance of 25 meters in 5.7 seconds, calculate

(a) The horizontal force required in newtons
(b) The work done in joules
(c) The work power in watts

Solution

   (a) horizontal force = 175 × 9.81 × 0.32 = 552.5 newtons
   (b) work done = 552.5 × 25 = 13812 joules
   (c) power = 13812/5.7 = 2423 watts

CHAPTER SUMMARY

□ F (force in newtons) = M (mass in kilograms) × A (acceleration in meters per second per second)
□ 1 newton = 100,000 dynes
□ Work done (joules) = F (force in newtons) × D (distance in meters)
□ 1 joule = 1 watt-second = 10,000,000 ergs
□ kWh or 1 kilowatt-hour = 3,600,000 joules
□ Torque (newton-meters) = F (force in newtons) × D (distance in meters)
□ 1 calorie = 4.18 joules
□ Power (watts) = joules per second
□ 1 horsepower = 746 watts

2
Electrical Units and Ohm's Law

IN THIS CHAPTER you will learn:

☐ The electrical units of current, charge, electromotive force, and power.

☐ The meaning of Ohm's Law as it applies to the concept of resistance.

☐ About practical resistors, some of which are identified by a color code.

☐ About the property of conductance, which is the reciprocal of resistance.

☐ The factors on which the resistance of a cylindrical conductor depends.

☐ How the resistance of a conductor varies with its temperature.

In a system of electrical units, it is necessary to define a fundamental unit and then build on this foundation to establish other units. In the rationalized mks or S.I. system (Systéme Internationale des Unités), the fundamental unit of current was adopted internationally in 1948. An electric current is the rate at which a quantity of electricity or charge flows past a given point in an electrical circuit. The unit of current is called the *ampere,* which commemorates a famous French scientist, André-Marie Ampére (1775–1836).

UNITS OF CURRENT AND CHARGE

The value of the ampere, commonly abbrevated amp or "A," is defined by an effect that occurs in electromagnetism. It is that value of current which, when flowing in each of two indefinitely long parallel conductors whose centers are separated by 1 meter in a vacuum, causes a force of 2×10^{-7} newtons per meter length to be exerted on each conductor. The apparatus that is needed to measure a current accurately from this definition is so very expensive and complex that it is normally not available outside of national laboratories.

The unit quantity of electricity or unit of charge is called the *coulomb,* which is named after another French scientist, Charles A. Coulomb. When a current of 1 ampere is maintained for 1 second, a charge of one coulomb flows past a given point in the circuit. It therefore follows that if a current of I amperes is maintained at a constant level for t seconds, the corresponding charge of Q coulombs is given by the following formula: Q (coulombs, abbreviated to C) = I (amperes) $\times$ t (seconds).

Consequently,

$$I \text{ (amperes)} = \frac{Q \text{ (coulombs)}}{t \text{ (seconds)}}$$

Note that with the aid of the time unit we have derived the coulomb from the ampere.

A charge of 1 coulomb is equivalent to the total negative charge possessed by 6.24×10^{18} electrons. This means that the charge carried by a single electron is $1/(6.24 \times 10^{18}) = 1.6019 \times 10^{-19}$ coulomb, while its mass is 9.1066×10^{-31} kilogram.

ELECTROCHEMICAL EQUIVALENT

When a current of 1 ampere passes between two copper plates immersed in an electrolyte of copper sulphate solution (a copper voltameter), copper is deposited on the negative plate at the rate of 0.0000003294 kilograms per second. The mass of 0.0000003294 kilograms is therefore liberated from the electrolyte by 1 coulomb. In a silver voltameter (two silver plates immersed in a silver nitrate solution), 1 coulomb liberates 0.0000011182 kilograms of silver. The amount of electrolyte liberated by 1 coulomb is called the *electrochemical equivalent.* Therefore the electrochemical equivalents of copper and silver are 3.294×10^{-7} kilograms per coulomb (kg/C) and 1.1182×10^{-6} kilograms per coulomb. For nickel and zinc, the electrochemical equivalents are respectively 3.04×10^{-7} and 3.38×10^{-7} kg/C.

If z is the electrochemical equivalent of a substance in kilograms per coulomb and I is the current in amperes maintained over an interval of t seconds, then the mass of the substance liberated = zIt kilograms.

Because 1 ampere is 1 coulomb per second, the unit of charge also can be called an ampere-second. A larger unit is the ampere-hour (Ah), which is equivalent to $60 \times 60 = 3600$ coulombs. For example, if a battery delivers a current of 3 amperes for 5 hours, the charge lost is 3A $\times$ 5h = 15 Ah, or $15 \times 3600 = 54,000$ coulombs.

Example 2-1

If a steady current of 12 A is maintained at a point for a time of 5 minutes, calculate the charge flowing past that point in (a) coulombs and (b) ampere-hours.

Solution

$$\begin{aligned}
\text{charge} &= \text{I (amperes)} \times \text{t (seconds)} \\
&= 12 \times 5 \times 60 \\
&= 3600 \text{ coulombs} \\
&= \frac{3600}{3600} \text{ ampere-hour} \\
&= 1 \text{ ampere-hour}
\end{aligned}$$

Example 2-2

A charge of 750 coulombs flows past a point in 5 minutes. Calculate the current in amperes.

Solution

$$\text{current} = \text{Q (coulombs)/t (seconds)} = 750/(5 \times 60) = 2.5\text{A}$$

Example 2-3

A constant current of 7A flows through a copper sulphate solution for a time interval of 50 minutes. Calculate the mass of copper liberated.

Solution

$$\begin{aligned}
\text{mass of copper liberated} &= zIt \\
&= 3.294 \times 10^{-7} \times 7 \times 50 \times 60 \\
&= 6.92 \times 10^{-3} \text{ kilogram}
\end{aligned}$$

UNIT OF ELECTROMOTIVE FORCE AND POTENTIAL DIFFERENCE

A source of electrical energy such as a battery provides an electromotive force (EMF) that impels electricity through a conductor connected across the battery terminals. The EMF is then balanced by an equal potential difference (p.d.) or difference of potential across the ends of the conductor. For both EMF and potential difference, the unit is the volt (V), named after the inventor of the first electrical battery, the Italian Count Alessandro Volta (1745 – 1827). In Fig. 2-1, E represents the electromotive force and V is the potential difference.

When a charge is driven between two points possessing a potential difference, work must be done. The volt can therefore be defined in terms of the coulomb and the joule. There is no need to redefine the joule because it already has been established during our derivation of the mechanical units. The relationship is: work done (joules) = potential difference (volts) × charge (coulombs), or

$$\text{potential difference (volts)} = \frac{\text{work done (joules)}}{\text{charge (coulombs)}}$$

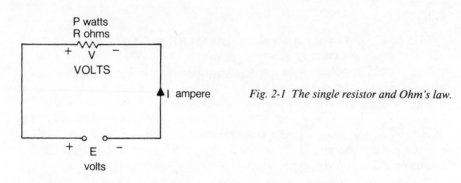

Fig. 2-1 The single resistor and Ohm's law.

In chapter one, you saw that

$$power \; (watts) = joules \; per \; second$$
$$= volts \times coulombs \; per \; second$$
$$= volts \times amperes$$

Consequently

$$power, \; P = IV \; watts$$
$$current, \; I = P/V \; amperes$$
$$voltage, \; V = P/I \; volts$$

The power in watts therefore is equal to the result of multiplying the current in amperes by the EMF or potential difference in volts.

Example 2-4

A charge of 5 coulombs moves through a potential difference of 15 volts in 3 seconds. Calculate the work done and the power involved.

Solution

$$work \; done = 15 \; volts \times 5 \; coulombs$$
$$= 75 \; joules$$
$$power = 75 \; joules/3 \; seconds = 25 \; watts$$

Example 2-5

A current of 6 amperes is associated with a potential difference of 75 volts for a period of 3 hours. Calculate the power and the total work in kilowatt-hours.

Solution

$$power = 6 \; amperes \times 75 \; volts$$
$$= 450 \; watts$$
$$total \; work = 450 \; watts \times 3 \; hours$$
$$= 1350 \; watt\text{-}hours$$
$$= 1.35 \; kWh$$

OHM'S LAW AND THE RESISTOR COLOR CODE

In 1827 Dr. George Simon Ohm discovered that under constant physical conditions, the current through a conductor was directly proportional to the difference of potential across the conductor. In other words, tripling the voltage tripled the current, and halving the voltage halved the current. Looked at another way, the ratio of the potential difference in volts to the current in amperes is a constant, and this relationship is known as *Ohm's Law* of constant proportionality. The constant is a measure of the conductor's opposition to current flow and is called the resistance R. The unit of resistance is the *ohm,* and its symbol is the Greek letter Ω (omega).

Referring to Fig. 2-1, Ohm's Law in equation form is

$$\frac{V}{I} = R$$

This leads to

$$V = I \times R \text{ and } I = \frac{V}{R}$$

Also

$$P = IV = \frac{V}{R} \times V = \frac{V^2}{R} \text{ watts, or } P = IV = I \times I \times R = I^2R \text{ watts}$$

Summarizing, the twelve relationships involving V, I, R and P are:

$$V = IR = \frac{P}{I} = \sqrt{P \times R} \text{ volts}$$

$$I = \frac{V}{R} = \frac{P}{V} = \sqrt{\frac{P}{R}} \text{ amps}$$

$$R = \frac{V}{I} = \frac{E}{I^2} = \sqrt{\frac{P}{I}} \text{ ohms}$$

$$P = IV = I^2R = \frac{V^2}{R} \text{ watts}$$

Most resistors used in electronics have their ohmic value and tolerance color-coded on the body of the resistor. The Electronics Industry Association code appears in Table 2-1.

Referring to Fig. 2-2, bands 1 and 2 indicate the first two significant figures of the rated value. Band 3 is the multiplier, and band 4 is the tolerance. For example, a resistor with band 1 red, band 2 violet, band 3 orange, and band 4 silver has a rated value of $27 \times 10^3 = 27,000 \ \Omega = 27 \text{ k}\Omega$, with a 10% tolerance. The permitted limits for this resistor are:

$$27,000 + \frac{27,000 \times 10}{100} = 29,700 \ \Omega = 2.97 \text{ k}\Omega,$$

$$\text{and } 27,000 - \frac{27,000 \times 10}{100} = 24,300 \ \Omega = 2.43 \text{ k}\Omega.$$

Table 2-1. EIA Resistor Color Code.

COLOR	BAND 1 (First Significant Figure)	BAND 2 (Second Significant Figure)	BAND 3 (Multiplier)	BAND 4 (Tolerance, Percent)
Black	0	0	$\times 10^0$	
Brown	1	1	$\times 10^1$	
Red	2	2	$\times 10^2$	
Orange	3	3	$\times 10^3$	
Yellow	4	4	$\times 10^4$	
Green	5	5	$\times 10^5$	
Blue	6	6	$\times 10^6$	
Violet	7	7	$\times 10^7$	
Gray	8	8	$\times 10^6$	
White	9	9	$\times 10^9$	
None				20
Silver			$\times 10^{-2}$	10
Gold			$\times 10^{-1}$	5

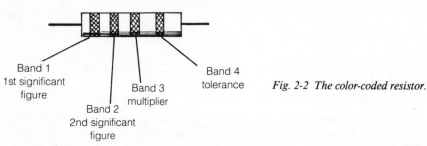

Band 1
1st significant
figure

Band 2
2nd significant
figure

Band 3
multiplier

Band 4
tolerance

Fig. 2-2 The color-coded resistor.

The common wattage ratings for such resistors are ⅛, ¼, ½, 1, and 2 watts.

Resistors are manufactured commercially only in certain ohmic values (see Table 2-2). The values as shown can be followed either by the appropriate number of zeros, or can be divided by 10 or 100 when gold or silver respectively appears in the third band.

The tolerance is the reason for using standard resistor values. For example, the permitted upper limit of a 220 Ω, 20% resistor is

$$220 + \frac{220 \times 20}{100} = 264 \ \Omega$$

The next highest rated value is 330 Ω, which would have a permitted lower limit of

$$330 - \frac{330 \times 20}{100} = 264 \ \Omega$$

The values are chosen therefore to avoid overlap between the tolerance spreads of resistors with adjacent rated values.

5 Percent Tolerance	10 Percent Tolerance	20 Percent Tolerance
10	10	10
11		
12	12	
13		
15	15	15
16		
18	18	
20		
22	22	22
24		
27	27	
30		
33	33	33
36		
39	39	
43		
47	47	47
51		
56	56	
62		
68	68	68
75		
82	82	
91		

Table 2-2. Standard Resistor Values

Example 2-6

The first three bands of a color-coded resistor are yellow, violet, and orange. Assuming that the resistance is the same as its rated value, what is the magnitude of the voltage applied across the resistor if the current flowing is 435 microamperes?

Solution

The rated value of the resistor is 4 (yellow), 7 (violet), 000 (orange) ohms, or $47 \text{ k}\Omega = 47 \times 10^3 \ \Omega$. The current is 435 microamperes $= 435 \times 10^{-6}$ A.

$$E = I \times R$$
$$= 435 \times 10^{-6} \times 47 \times 10^3 \text{ V}$$
$$= 4.35 \times 10^2 \times 10^{-6} \times 4.7 \times 10^1 \times 10^3$$
$$= 4.35 \times 4.7$$
$$= 20.4 \text{ V}$$

Example 2-7

The current flowing through a particular 1-watt resistor is 23.5 mA. What is the highest value of voltage that can be applied across the resistor without exceeding its power rating?

Solution

The current is 23.5 mA = 23.5×10^{-3} A.

$$E = \frac{P}{I} = \frac{1}{23.5 \times 10^{-3}} = \frac{1 \times 10^3}{23.5} = \frac{1000}{23.5} = 42.6 \text{ volts}$$

Example 2-8

The first three bands of a ½-watt color-coded resistor are green, blue, and yellow. What is the value of the voltage that can be applied across the resistor without exceeding its power rating?

Solution

The value of the resistor is 5 (green), 6 (blue), 0000 (yellow) ohms = 560 kΩ.

$$\begin{aligned} E = \sqrt{P \times R} &= \sqrt{½ \times 560{,}000} \\ &= \sqrt{280{,}000} \\ &= \sqrt{28 \times 10^4} = \sqrt{28} \times 10^2 = 529 \text{ V} \end{aligned}$$

Example 2-9

A 12V battery is connected across a color-coded resistor whose first three bands are red, red, and green. Assuming the resistance is the same as its rated value, what is the value of the current flowing through the resistor?

Solution

The value of the resistor is 2 (red), 2 (red), 00000 (green) ohms = 2200 kΩ = 2.2 MΩ.

$$I = \frac{E}{R} = \frac{12}{2.2 \times 10^6} = \frac{12 \times 10^{-6}}{2.2} A = 5.5 \ \mu A$$

Example 2-10

A 25-watt electric light bulb operates from a 115V dc source. What is the value of the current that is flowing in the circuit?

Solution

$$\text{current, } I = \frac{P}{E} = \frac{25}{115} = 0.217 \text{ A} = 217 \text{ mA}$$

Example 2-11

The first three bands of a ¼-watt color-coded resistor are blue, grey, and brown. What is the highest value of current that can flow through the resistor without exceeding its power rating (assuming that the actual resistance is the same as the color-coded value)?

Solution

The value of the resistor is 6 (blue), 8 (grey), 0 (brown), or 680 Ω. Then:

$$\sqrt{\frac{P}{R}} = \sqrt{\frac{\frac{1}{4}}{680}} = \sqrt{\frac{0.25}{680}} = \sqrt{0.0003676}$$
$$= \sqrt{3.676 \times 10^{-4}} \text{ A}$$
$$= 1.9 \times 10^{-2} \text{ A}$$
$$= 19 \text{ mA}$$

Example 2-12

When 15.4V is applied across a resistor, the current is measured as 38 μA. What is the value of the resistor?

Solution

$$\text{resistance, R} = \frac{E}{I} = \frac{15.4}{38 \times 10^{-6}}$$
$$= \frac{15.4 \times 10^6}{38}$$
$$= 0.405 \times 10^6 \ \Omega = 405 \text{ k} \ \Omega$$

Example 2-13

A 75-watt electric light bulb is designed to operate from a 110 V dc supply. What is the "hot" resistance of the bulb's filament?

Solution

$$\text{resistance, R} = \frac{E^2}{P} = \frac{110^2}{75} = \frac{110 \times 110}{75} = 161 \ \Omega$$

Example 2-14

When a current of 237 mA flows through a wire-wound resistor, the power dissipated is 55 W. What is the value of the resistance?

Solution

$$\text{resistance, R} = \frac{P}{I^2} = \frac{55}{(237 \times 10^{-3})^2}$$
$$= \frac{55}{(2.37 \times 10^{-1})^2}$$
$$= \frac{55}{2.37 \times 2.37 \times 10^{-2}}$$
$$= \frac{55 \times 10^2}{2.37 \times 2.37} = 979 \ \Omega$$

Example 2-15

When a 220 V dc source is applied across a resistive load, the current taken from the source is 17.3 A. How much power is dissipated in the load?

Solution

$$\text{power, } P = E \times I = 220 \times 17.3 = 3810 \text{ W} = 3.81 \text{ kW}$$

Example 2-16

A current of 23.7 mA flows through a 2-watt, 3.3 kΩ resistor. How much power is dissipated in the resistor?

Solution

$$
\begin{aligned}
\text{power, } P = I^2 \times R &= (23.7 \times 10^{-3})^2 \times 3.3 \times 10^3 \\
&= (2.37 \times 10^{-2})^2 \times 3.3 \times 10^3 \\
&= 2.37 \times 2.37 \times 3.3 \times 10^{-4} \times 10^3 \\
&= 1.85 \text{ W}
\end{aligned}
$$

Example 2-17

The first three bands of a color-coded 2-watt resistor are grey, red, black. If this resistor is connected across a 9 V battery, what is the amount of the power dissipated?

Solution

The value of the resistor is 8 (grey), 2 (red) ohms (the multiplier is black and therefore no zeros appear after the 2). Then

$$P = \frac{E^2}{R} = \frac{9^2}{82} = \frac{81}{82} = 0.99 \text{ W}$$

CONDUCTANCE

The resistance of a conductor is a measure of its opposition to current flow and is defined from the relationship, $R = V/I$. By contrast, conductance measures the ease with which current will flow through a conductor; conductance, G, is therefore the inverse or reciprocal of resistance and is defined by

$$G = \frac{I}{V} = \frac{1}{R}$$

In the SI system, the unit of conductance is the *siemens* (S). Prior to the introduction of the siemens, the unit of conductance was the *mho* (ohm spelled backwards).

Example 2-18

The first three bands of a color-coded resistor are red, red, gold. What is the resistor's conductance?

Solution

Gold or silver in the third band represent multiplier values of 0.1 and 0.01, respectively. The value of the resistor is therefore $22 \times 0.1 = 2.2 \ \Omega$. Its conductance is

$$G = \frac{1}{2.2} = 0.45 \text{ S}$$

Example 2-19

The total conductance of a dc circuit is 3.75×10^{-4} S. If the dc source voltage is 165 V, what is the amount of current taken from the source?

Solution

$$\begin{aligned}
I = E \times G &= 165 \times 3.75 \times 10^{-4} \\
&= 1.65 \times 10^2 \times 3.75 \times 10^{-4} \\
&= 6.1875 \times 10^{-2} \text{ A} \\
&= 61.9 \text{ mA}
\end{aligned}$$

Example 2-20

When an EMF of 12 V is applied across a resistive load, the dc load current is 250 mA. What is the conductance of the load?

Solution

$$\text{conductance, } G = \frac{I}{E} = \frac{250 \times 10^{-3}}{12} = 0.021 \text{ S}$$

RESISTANCE OF THE CYLINDRICAL CONDUCTOR

The resistance of the cylindrical conductor is directly proportional to the conductor's length, inversely proportional to its cross-sectional area, and depends on the material from which the conductor is made. In equation form

$$R = \rho \times \frac{L}{A}$$

where:

R = resistance of the conductor in ohms
L = length of the conductor in meters

A = cross-sectional area of the conductor in square meters
ρ (rho) = specific resistance or resistivity of the conductor's material

With L in meters and A in square meters, the value of ρ will be in SI units. A conductor one meter long with a cross-sectional area of 1 square meter has a resistance of ρ ohms. The SI unit of resistivity is the ohm meter ($\Omega \cdot m$), but because the resistance of a conductor depends on the temperature, the value of the resistivity is quoted for a particular temperature, which is normally 20°C (see Table 2-3).

Table 2-3. Relative Resistivity of Various Conductors.

Conductor's Material	Resistivity ($\Omega \cdot m$) at 20°C
Silver	1.64×10^{-8}
Copper (annealed)	1.724×10^{-8}
Aluminum	2.83×10^{-8}
Tungsten	5.5×10^{-8}
Nickel	7.8×10^{-8}
Iron (pure)	12.0×10^{-8}
Constantan	49.0×10^{-8}

The cgs unit of resistivity is the ohm·centimeter (1 ohm·meter = 100 ohm·centimeters), but in the British system, ρ is measured in ohm·circular mil per foot, where one circular mil (cmil) is the area of a circle whose diameter is 1 linear mil = 1/1000 inch. The use of circular mils avoids the need for using π when calculating the cross-sectional area. If, for example, the diameter is 3 mils, the area is $3^2 = 9$ cmils. In this system of units, the resistance of the cylindrical conductor is:

$$R = \rho \frac{L}{A} = \rho \frac{L}{d^2}$$

where:

R = resistance of the conductor in ohms
ρ = specific resistance in ohm·cmil/ft
L = length of the conductor in feet
A = cross-sectional area in cmils
d = diameter in linear mils

The conversion factor between the SI and British systems is 1 ohm·meter = 6.015×10^8 ohm circular mil per foot, so that for annealed copper, $\rho = 1.7 \times 10^{-8}\ \Omega \cdot m = 1.724 \times 10^{-8} \times 6.015 \times 10^8$ or 10.37 ohm·cmil/ft.

Just as conductance is the reciprocal of resistance, conductivity (σ, or sigma) is the reciprocal of resistivity. The conductivity is the conductance of a meter length of the material with a cross sectional area of 1 square meter.

Therefore,

$$\sigma = \frac{1}{\rho}$$

and is measured in siemens per meter (S/m). For example, σ for iron at 20°C is $1/(12.0 \times 10^{-8}) = 8.33 \times 10^7$ S/m.

Example 2-21

An electrical conductor 2.5 meters long has a cross-sectional area of 0.75 square millimeter and a resistance of 0.012 ohm. What is the resistance of 30 meters of wire that is made from the same material and has a cross-sectional area of 0.4 square millimeter?

Solution

The resistance, R, is directly proportional to the conductor's length, L, and inversely proportional to the cross-sectional area, A. Therefore,

$$\frac{new\ R}{old\ R} = \frac{new\ L/new\ A}{old\ L/old\ A}$$

$$\frac{new\ R}{0.012} = \frac{30/0.4}{2.5/0.75} = \frac{30 \times 0.75}{2.5 \times 0.4} = 22.5$$

$$new\ resistance = 22.5 \times 0.012 = 0.27\ \Omega$$

Example 2-22

What is the resistance of a nickel conductor that is 3 meters long and has a cross-sectional area of 2 square millimeters?

Solution

$$A = 2 \times 10^{-6}\ m^2,\ L = 3\ m,\ \rho = 7.8 \times 10^{-8}\ \Omega \cdot m$$

$$resistance,\ R = \rho \frac{L}{A} = 7.8 \times 10^{-8}\ \Omega \cdot m \times \frac{3m}{2 \times 10^{-6}m^2} = 0.117\ \Omega$$

Example 2-23

A 50-meter length of annealed copper wire has a circular cross-sectional area whose diameter is 0.85 mm. What is its resistance at 20°C?

Solution

$$cross\text{-}sectional\ area,\ A = \frac{\pi d^2}{4} = \frac{\pi}{4} \times (0.85 \times 10^{-3})^2$$

$$= 5.67 \times 10^{-7}\ m^2$$

$$L = 50\ m,\ \rho = 1.724 \times 10^{-8}\ \Omega \cdot m$$

$$resistance\ R = \rho \frac{L}{A} = 1.724 \times 10^{-8}\ \Omega \cdot m \times \frac{50\ m}{5.67 \times 10^{-7}}$$

$$= 1.52\ \Omega$$

Example 2-24

A tungsten filament has a length of 3 inches and a cross-sectional area of 2 circular mils. What is its resistance at 20°C?

Solution

$$L = 3/12 = 0.25 \text{ ft}; \rho = 5.5 \times 10^{-8} \times 6.015 \times 10^{8} = 33.1 \text{ ohm} \cdot \text{cmil/ft}$$

$$\text{resistance, } R = \rho \frac{L}{A} = 33.1 \times \frac{0.25}{2} = 4.14 \; \Omega$$

TEMPERATURE COEFFICIENT OF RESISTANCE

For most conducting materials such as copper and aluminum, the resistance rises in a linear manner with an increase of temperature over normal temperature ranges. In contrast, there are some alloys, for example, eureka (60% copper, 40% nickel), whose resistance is affected little by temperature. Finally, there are some elements (carbon, germanium, silicon) in which there is a reduction in resistance as the temperature increases.

The change in resistance with temperature is measured by the temperature coefficient of resistance, α (alpha). This commonly is defined as the ohmic increase per ohm of resistance at 20°C per degree centigrade (Celsius) rise in temperature. For example, α at 20°C for copper is $+0.00393 \; \Omega/\Omega/$°C; this means that if a length of copper wire has a resistance of one ohm at 20°C, the resistance increases by $0.00393 \; \Omega$ for every 1°C rise in temperature. The values of the temperature coefficient for various materials appear in Table 2-4; notice that constantan is an alloy with a very low temperature coefficient and that carbon has a negative coefficient. Also, see Table 2-5.

Conductor Material	$\alpha(20°C)$
Silver	0.0038
Aluminum	0.0039
Copper	0.00393
Tungsten	0.0045
Iron	0.0055
Nickel	0.006
Constantan	0.0000008
Carbon	−0.0005

Table 2-4. Resistance Temperature Coefficients

In equation form

$$R_{T}° = R_{20°C} (1 + \alpha_{20°C} (T° - 20°))$$
$$\text{or } R_{20°C} = R_{T°}/(1 + \alpha_{20°C} (T° - 20°))$$

where

$$R_{T°} = \text{conductor's resistance at } T°C.$$
$$R_{20°C} = \text{conductor's resistance at } 20°C.$$
$$\alpha = \text{temperature coefficient at } 20°C.$$

Table 2-5. Electrical Units

Quantity	Unit	Unit Symbol	Letter Symbol
Current	Ampere	A	I
Charge	Coulomb, Ampere-Hour	C,Ah	Q
EMF or p.d.	Volt	V	E V
Work, Energy	Joule, Watt-Hour, Kilowatt-Hour	J,Wh kWh	W
Power	Watt	W	P
Resistance	Ohm	Ω	R
Conductance	Siemens (MHO)	S	G
Specific Resistance, Resistivity	Ohm.Meter	Ωm	ρ
Conductivity	Siemens per Meter	S/m	σ
Temperature Coefficient of Resistance	Ohms per Degree Centgrade	$\Omega/\Omega/°C$	α

Because

$$R_{T°} = \rho_{20°C} \times \frac{L}{A}$$

$$R_{T°} = \rho_{20°C} \times \frac{L}{A}(1 + \alpha_{20°C}(T° - 20°))$$

and

$$\rho_{T°} = \rho_{20°C}(1 + \alpha_{20°C}(T° - 20°))$$

Example 2-25

A length of copper wire has a resistance of 25 ohms at 20°C. Calculate its resistance at (a) 100°C and (b) 0°C.

Solution

$$
\begin{aligned}
R_{100°C} &= 25(1 + 0.00393 \times (100° - 20°)) \\
&= 25 \times (1 + 0.00393 \times 80) \\
&= 25 \times 1.3144 \\
&= 32.86 \ \Omega \\
R_{0°C} &= 25(1 + 0.00393 \times (0° - 20°)) \\
&= 25(1 - 0.00393 \times 20) \\
&= 25 \times 0.9214 \\
&= 23.04 \ \Omega
\end{aligned}
$$

Example 2-26

Calculate the resistance of 250 meters of nickel wire with a cross-sectional area of 1.75 square millimeters at 60°C.

Solution

$$\rho_{20°C} = 7.8 \times 10^{-8}\ \Omega \cdot m,\ \alpha_{20°C} = 0.006,$$
$$A = 1.75 \times 10^{-6}\ \text{square meter}$$

$$\text{Then } R_{60°C} = \frac{7.8 \times 10^{-8} \times 250}{1.75 \times 10^{-6}} \times (1 + 0.006 \times (60° - 20°))$$

$$= \frac{7.8 \times 10^{-2} \times 250 \times 1.24}{1.75}$$

$$= 13.8\ \Omega$$

Example 2-27

What is the resistivity of iron at 80°C?

Solution

$$\rho_{20°C} = 12.0 \times 10^{-8}\ \Omega \cdot m,$$
$$\alpha_{20°C} = 0.0055$$
$$\rho_{80°C} = 12.0 \times 10^{-8}(1 + 0.0055 \times (80° - 20°))$$
$$= 12.0 \times 10^{-8} \times 1.33$$
$$= 16.0 \times 10^{-8}\ \Omega \cdot m$$

Example 2-28

A tungsten filament has a resistance of 25 Ω at 2000°C. What is its resistance at room temperature (20°C)?

Solution

$$\rho_{20°C} = 0.0045\ \Omega/\Omega/°C$$
$$R_{20°C} = 25/(1 + 0.0045 \times (2000 - 20))$$
$$= 25/(1 + 0.0045 \times 1980)$$
$$= 25/9.91$$
$$= 2.52\ \Omega$$

CHAPTER SUMMARY

☐ Q(coulombs) = I (amperes) × t (seconds)
☐ Mass of substance liberated from electrolyte
   = z (kg/C) × I (amperes) × t (seconds)
   = z (kg/C) × Q (coulombs)
☐ Electrical energy or work = Pt joules

☐ Current, $I = \dfrac{V}{R} = \dfrac{P}{V} = \sqrt{\dfrac{P}{R}}$ amperes

☐ Voltage, $V = \dfrac{P}{I} = IR = \sqrt{P \times R}$ volts

☐ Power, $P = V \times I = I^2 R = \dfrac{V^2}{R}$ watts

☐ Resistance, $R = \dfrac{V}{I} = \dfrac{P}{I^2} = \dfrac{V^2}{P}$ ohms

☐ Conductance, $G = \dfrac{1}{R} = \dfrac{I}{V} = \dfrac{I^2}{P} = \dfrac{P}{V^2}$ siemens

☐ Resistance of cylindrical conductor

$= \begin{matrix} \rho(\text{resistivity} \\ \text{in ohm} \cdot \text{meter}) \end{matrix} \times \dfrac{L \text{ (meters)}}{A \text{ (square meters)}}$

☐ Resistance of conductor at $T\,°C = R_{20°C}(1 + \alpha_{20°C}(T° - 20°))$

3
Resistor Networks

IN THIS CHAPTER you will specifically learn:

☐ How to distinguish a series circuit from other resistor arrangements.

☐ About the manner in which a current flows through a series circuit.

☐ How to determine the value of the individual voltages across resistors in series.

☐ How to calculate the value of the total equivalent resistance of a series circuit.

☐ How to determine the powers dissipated in the individual resistors and the total power in the series circuit.

☐ How to analyze a series circuit by using a step-by-step procedure.

☐ About using the voltage division rule to obtain the value of a particular voltage drop.

☐ About the use of voltage dividers and how they are formed by connecting resistors in series.

☐ About the use of ground as a common connection and a voltage reference level.

☐ How to obtain the correct voltage for a load when only a higher voltage is available.

☐ About the equality between the source voltage and the voltage drop across each of the resistors in parallel.

☐ That the sum of the parallel branch currents is equal to the total current drawn from the voltage source.

☐ About the concept of the total equivalent resistance of a parallel circuit and how its value can be determined from the reciprocal or product-over-sum formulas.

☐ How to calculate the value of the total equivalent conductance in a parallel circuit.

□ To understand the relationship between the total power and the power dissipated in the individual parallel resistors.

□ How to analyze a parallel circuit by using a step-by-step procedure.

□ To apply the current division rule in the analysis of a parallel circuit.

□ How open circuits affect a parallel arrangement of resistors.

□ How short circuits affect a parallel arrangement of resistors.

□ To use a step-by-step method for analyzing complex series–parallel networks of resistors.

Referring to Fig. 3-1, the resistors R1, R2 . . . . RN are connected in series because they are joined end-to-end in succession. The properties of this circuit are:

1. There is only a single path for current flow and the current has the same value throughout the circuit.

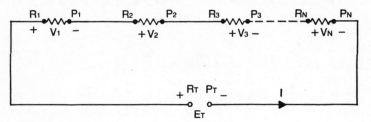

Fig. 3-1 Resistors in series.

RESISTORS IN SERIES

2. The sum of the individual voltage drops across the resistors is equal to the value of the source voltage. This is Kirchhoff's Voltage Law.

For N resistors in series, $E_T = V_1 + V_2 + V_3 + \cdots + V_N$, and $V_1 = IR_1$, $V_2 = IR_2 \ldots V_N = IR_N$.

Note that the highest value resistor develops the greatest voltage drop. If the N resistors all have the same resistance R, $E_T = NV$ and $V_1 = V_2 = \cdots = V_N = V = IR$, where V is the voltage drop across each resistor.

3. The total equivalent resistance is equal to the sum of the individual resistances.

For N resistors in series,

$$R_T = R_1 + R_2 + R_3 + \cdots + R_N$$

$$I = \frac{E_T}{R_T} = \frac{E_T}{R_1 + R_2 + R_3 + \cdots + R_N}$$

If the N resistors all have the same resistance R, $R_T = NR$ and $I = E_T/NR$.

4. The total power delivered from the source is equal to the sum of the individual powers dissipated in the resistors.

For N resistors in series,

$$P_T = P_1 + P_2 + P_3 + \cdots + P_N$$
$$= I^2R_T = \frac{E_T{}^2}{R_T} = E_TI.$$

where

$$P_1 = I^2R_1 = \frac{V_1{}^2}{R_1} = IV_1$$

etc.

Note that the highest value resistor dissipates the most power.
If the N resistors all have the same resistance, R,

$$P_T = NP \text{ and } P_1 = P_2 = \cdots = P_N = I^2R = \frac{V^2}{R} = IV$$

VOLTAGE-DIVISION RULE

5. The source voltage divides between the resistors in *direct* proportion
to the values of the resistors (voltage division rule). Then

$$V_1 = V_2 \times \frac{R_1}{R_2} = E_T \times \frac{R_1}{R_T}$$

etc.

For two resistors in series,

$$V_1 = E_T \times \frac{R_1}{R_1 + R_2}$$

and

$$V_2 = E_T \times \frac{R_2}{R_1 + R_2}$$

The total power delivered from the source also is divided between the
resistors in direct proportion to the values of the resistors.

$$P_1 = P_2 \times \frac{R_1}{R_2} = P_1 \times \frac{R_1}{R_T}$$

etc.

THE OPEN CIRCUIT

6. An open circuit has theoretically infinite resistance. No current can
flow through an open circuit, but in a series string of resistors, the source
voltage appears across the open circuit.

THE SERIES-DROPPING RESISTOR

7. A series, dropping resistor can be used to reduce a source voltage down to the level required by a load. Referring to Fig. 3-2,

$$V_D = E - V_L = E - I_L \times R_L$$

$$R_D = \frac{V_D}{I_L}$$

$$P_D = I_L \times V_D$$

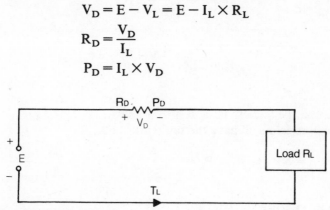

Fig. 3-2 The series-dropping resistor.

Example 3-1

In Fig. 3-3, what are the values of the voltages V_{A-B}, V_{D-E}, V_{B-H}, V_{H-P}, V_{N-S}, V_{E-P}, V_{A-S}, and V_{Q-J}? (V_{A-B} means the voltage between points A and B).

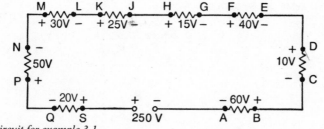

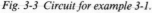

Fig. 3-3 Circuit for example 3-1.

Solution

$V_{A-B} = 60$ V

$V_{D-E} = 0$ V (Between D and E there is only a connecting wire across which there is a negligible voltage drop)

$V_{B-H} = V_{B-c} + V_{C-D} + V_{D-E} + V_{E-F} + V_{F-G} + V_{G-H}$
$\quad = 0$ V $+ 10$ V $+ 0$ V $+ 40$ V $+ 0$ V $+ 15$ V $= 65$ V

$V_{H-P} = V_{H-J} + V_{J-K} + V_{K-L} + V_{L-M} + V_{M-N} + V_{N-P}$
$\quad = 0$ V $+ 25$ V $+ 0$ V $+ 30$ V $+ 0$ V $+ 50$ V $= 105$ V

$V_{N-S} = V_{N-P} + V_{P-Q} + V_{Q-S}$
$\quad = 50$ V $+ 0$ V $+ 20$ V $= 70$ V

$V_{E-P} = V_{E-F} + V_{F-G} + V_{G-H} + V_{H-J} + V_{J-K} + V_{K-L} + V_{L-M} + V_{M-N} + V_{N-P}$
$\quad = 40$ V $+ 0$ V $+ 15$ V $+ 0$ V $+ 25$ V $+ 0$ V $+ 30$ V $+ 0$ V $+ 50$ V $= 160$ V

$V_{A-S} = 250$ V (The 250 V battery is connected between the points A and S)

$V_{Q-J} = V_{Q-P} + V_{P-N} + V_{N-M} + V_{M-L} + V_{L-K} + V_{K-J}$
$= 0$ V $+ 50$ V $+ 0$ V $+ 30$ V $+ 0$ V $+ 25$ V $= 105$ V

Example 3-2

In Fig. 3-4, what are the values of the potential differences V_{A-B}, V_{B-E}, V_{C-F}, V_{A-D}, and V_{D-F}?

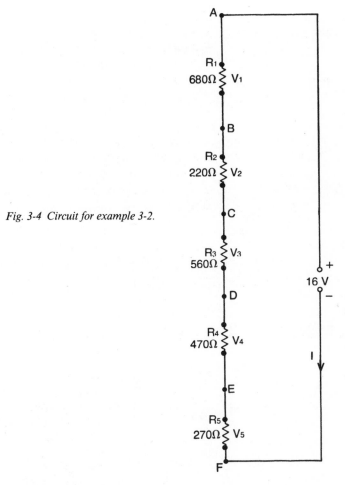

Fig. 3-4 Circuit for example 3-2.

Solution

$$\text{total resistance, } R_T = 680 + 220 + 560 + 470 + 270$$
$$= 2200 \ \Omega$$

$$\text{current, } I = \frac{E_T}{R_T} = \frac{16 \text{ V}}{2200 \ \Omega} = 0.0072727 \text{ A}$$

$$\text{potential difference, } V_1 = IR_1 = 0.0072727 \text{ A} \times 680 \ \Omega$$
$$= 4.945436 \text{ V}$$

$$V_2 = IR_2 = 0.0072727 \text{ A} \times 220 \text{ } \Omega$$
$$= 1.599994 \text{ V}$$
$$V_3 = IR_3 = 0.0072727 \text{ A} = 560 \text{ } \Omega$$
$$= 4.072712 \text{ V}$$
$$V_4 = IR_4 = 0.0072727 \text{ A} = 470 \text{ } \Omega$$
$$= 3.418169 \text{ V}$$
$$V_5 = IR_5 = 0.0072727 \text{ A} = 270 \text{ } \Omega$$
$$= 1.963629 \text{ V}$$

Voltage check:

$$\text{sum of voltage drops} = V_1 + V_2 + V_3 + V_4 + V_5$$
$$= 4.945436 + 1.599994 + 4.072712$$
$$+ 3.418169 + 1.963629$$
$$= 15.99994 \text{ V}$$

This compares with the applied voltage of 16 volts; the small difference is due to the rounding off.

$$V_{A-B} = V_1 = 4.945436 = 4.95 \text{ V, rounded off}$$
$$V_{B-E} = V_2 + V_3 + V_4 = 9.090875 = 9.09 \text{ V, rounded off}$$
$$V_{C-F} = V_3 + V_4 + V_5 = 9.45451 = 9.45 \text{ V, rounded off}$$
$$V_{A-D} = V_1 + V_2 + V_3 = 10.618142 = 10.62 \text{ V, rounded off}$$
$$V_{D-F} = V_4 + V_5 = 5.381798 = 5.38 \text{ V, rounded off}$$

Example 3-3

Find the values of V_1, V_2, V_3, V_4, V_5, P_1, P_2, P_3, P_4, P_5 and the total power, P_T, dissipated in the circuit of Fig. 3-5.

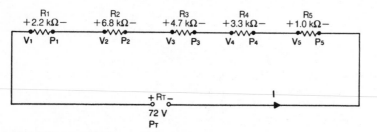

Fig. 3-5 Circuit for example 3-3.

Solution

$$\text{total resistance, } R_T = 2.2 + 6.8 + 4.7 + 3.3 + 1.0$$
$$= 18 \text{ k}\Omega$$

$$\text{circuit current, } I = \frac{E_T}{R_T} = \frac{72 \text{ V}}{18 \text{ k}\Omega} = 4 \text{ mA}$$

The individual voltage drops are:

$$V_1 = IR_1 = 4 \text{ mA} \times 2.2 \text{ k}\Omega = 8.8 \text{ V}$$

$$V_2 = IR_2 = 4 \text{ mA} \times 6.8 \text{ k}\Omega = 27.2 \text{ V}$$
$$V_3 = IR_3 = 4 \text{ mA} \times 4.7 \text{ k}\Omega = 18.8 \text{ V}$$
$$V_4 = IR_3 = 4 \text{ mA} \times 3.3 \text{ k}\Omega = 13.2 \text{ V}$$
$$V_5 = IR_5 = 4 \text{ mA} \times 1.0 \text{ k}\Omega = 4.0 \text{ V}$$

The sum of the voltage drops is $V_1 + V_2 + V_3 + V_4 + V_5 = 8.8 + 27.2 + 18.8 + 13.2 + 4.0 = 72.0$ V, which exactly balances the source voltage, E_T.

The individual powers dissipated in the resistors are

$$P_1 = IV_1 = 4 \text{ mA} \times 8.8 \text{ V} = 35.2 \text{ mW}$$
$$P_2 = IV_2 = 4 \text{ mA} \times 27.2 \text{ V} = 108.8 \text{ mW}$$
$$P_3 = IV_3 = 4 \text{ mA} \times 18.8 \text{ V} = 75.2 \text{ mW}$$
$$P_4 = IV_4 = 4 \text{ mA} \times 13.2 \text{ V} = 52.8 \text{ mW}$$
$$P_5 = IV_5 = 4 \text{ mA} \times 4.0 \text{ V} = 16.0 \text{ mW}$$

The total power dissipated by the resistors is

$$P_T = P_1 + P_2 + P_3 + P_4 + P_5$$
$$= 35.2 + 108.8 + 75.2 + 52.8 + 16.0 = 288.0 \text{ mW}$$

The total power delivered by the source is

$$E_T \times I = 72 \text{ V} \times 4 \text{ mA} = 288 \text{ mW}$$

which exactly balances the value of P_T.

Example 3-4

In Fig. 3-6, calculate the values of V_1, E_T, and P_3.

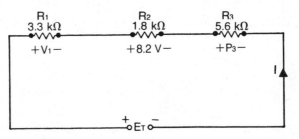

Fig. 3-6 Circuit for example 3-4.

Solution

$$V_1 = V_2 \times \frac{R_1}{R_2} = 8.2 \times \frac{3.3}{1.8} = 15.03 \text{ V}$$

$$E_T = V_1 \times \frac{R_T}{R_2} = 8.2 \times \frac{(3.3 + 1.8 + 5.6)}{1.8} = 8.2 \times \frac{10.7}{1.8} = 48.74 \text{ V}$$

$$P_3 = P_2 \times \frac{R_3}{R_2}$$

where

$$P_2 = \frac{V_2^2}{R_2} = \frac{(8.2 \text{ V})^2}{1.8 \text{ k}\Omega} = 37.36 \text{ mW}$$

$$P_3 = 37.36 \times \frac{5.6}{1.8} = 116.2 \text{ mW}$$

Example 3-5

Using Fig. 3-7, calculate the values of V_2 and V_3.

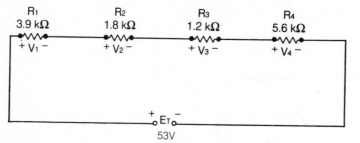

Fig. 3-7 *Circuit for example 3-5.*

Solution

$$V_2 = E_T \times \frac{R_2}{R_T} = 53 \times \frac{1.8}{3.9 + 1.8 + 1.2 + 5.6}$$

$$= 53 \times \frac{1.8}{12.5} = 7.63 \text{ V}$$

$$V_3 = E_T \times \frac{R_3}{R_T} = 53 \times \frac{1.2}{12.5} = 5.09 \text{ V}$$

V_3 also could be calculated from

$$V_3 = V_2 \times \frac{R_3}{R_2} = 7.63 \times \frac{1.2}{1.8} = 5.09 \text{ V}$$

Example 3-6

In Fig 3-8, what is the voltage between the points X and Y?

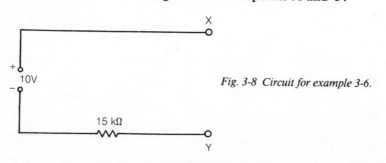

Fig. 3-8 *Circuit for example 3-6.*

Solution

Because there is an open circuit between X and Y, there is no current flow through the 15 kΩ resistor and therefore no voltage drop across this resistor. Consequently the voltage between X and Y is 10 V.

Example 3-7

A relay whose coil resistance is 40 Ω is designed to function correctly when 2 A flows through the coil. If the relay operates from 115 V dc, find the value of the series-dropping resistor required and calculate its power dissipation.

Solution

Referring to Fig. 3-2, $R_L = 40\ \Omega$, $I_L = 2$ A, and E = 115 V.

$V_D = E - I_L R_L = 115\ V - 2\ A \times 40\ \Omega = 115\ V - 80\ V = 35\ V$
$R_D = V_D/I_L = 35\ V/2\ A = 17.5\ \Omega$
$P_D = I_L \times V_D = 2\ A \times 35\ V = 70\ W$

GROUND AS A RETURN LINE
AND AS A VOLTAGE REFERENCE LEVEL

Ground can be considered as any large mass of conducting material, (for example the metal chassis of a transmitter) so that there is essentially zero resistance between any two ground points. The main reason for using a ground system is to simplify circuitry by saving on the amount of wiring required. Ground is then used as the return path for many circuits so that at any time there are a number of currents flowing through the ground system. However because of the ground's zero resistance property there is no voltage developed between any two ground points and therefore there is no interference between the various circuits. This is illustrated in Figs. 3-9A and 3-9B, which show two circuits with four connecting wires. However, with a common ground as the return path (Fig. 3-9C), only two connecting wires are needed, and there is zero voltage drop between the ground points.

In addition to providing a return path, ground is also chosen as a reference level of zero volts, and the voltage or potential at any point can then be measured relative to ground. In Fig. 3-10A, point P is 10 volts positive with respect to point Q and point Q is 10 volts negative with respect to point P. However if point Q is grounded at zero volts (Fig. 3-10B), point P carries a voltage of positive 10 volts (+ 10 V), while if P is grounded (Fig. 3-10C), the potential at Q is negative 10 volts (− 10 V). Individual points therefore can either possess a positive or a negative potential with respect to ground. If the electron flow is from ground to a particular point, that point's potential will be positive, but if the flow is reversed, the potential is negative.

In any particular circuit, the potentials at the various points depend on which point is grounded; however, the voltage drops between any two points remain the same.

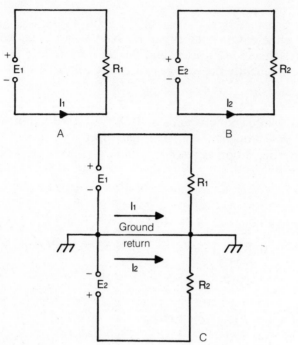

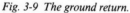

Fig. 3-9 The ground return.

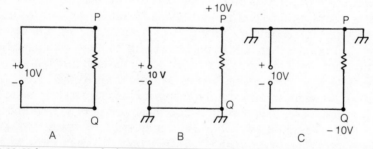

Fig. 3-10 Voltage measured relative to ground.

Example 3-8

In the circuit of Fig. 3-11, each of the points is grounded in turn. What voltages with respect to ground then exist at all of the other points in the circuit?

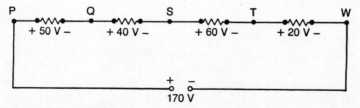

Fig. 3-11 Circuit for example 3-8.

Solution

P grounded

> potential at Q $= -50$ V
> potential at S $= -(50 + 40) = -90$ V
> potential at T $= -(50 + 40 + 60) = -150$ V
> potential at W $= -(50 + 40 + 60 + 20) = -170$ V

Q grounded

> potential at P $= +50$ V
> potential at S $= -40$ V
> potential at T $= -(40 + 60) = -100$ V
> potential at W $= -(40 + 60 + 20) = -120$ V

S grounded

> potential at P $= +(40 + 50) = +90$ V
> potential at Q $= +40$ V
> potential at T $= -60$ V
> potential at W $= -(60 + 20) = -80$ V

T grounded

> potential at P $= +(60 + 40 + 50) = 150$ V
> potential at Q $= +(60 + 40) = +100$ V
> potential at S $= +60$ V
> potential at W $= -20$ V

W grounded

> potential at P $= +(20 + 60 + 40 + 50) = +170$ V
> potential at Q $= +(20 + 60 + 40) = +120$ V
> potential at S $= +(20 + 60) = 80$ V
> potential at T $= +20$ V

RESISTORS IN PARALLEL

Referring to Fig. 3-12A, the resistors $R_1, R_2 \ldots R_N$ are connected in parallel because they are joined between two common points X and Y. Electronically this is the same as connecting the resistors between two common lines as shown in Fig. 3-12B. The properties of this circuit are:

1. The voltage drop across each resistor is the same and equal to the source voltage, or

$$E_T = V_1 = V_2 = \cdots = V_N$$

2. The sum of the individual branch currents flowing through the resistors is equal to the total current drawn from the source voltage (*Kirchhoff's Current Law*).

For N resistors is parallel:

$$I_T = I_1 + I_2 + I_3 + \cdots + I_N$$

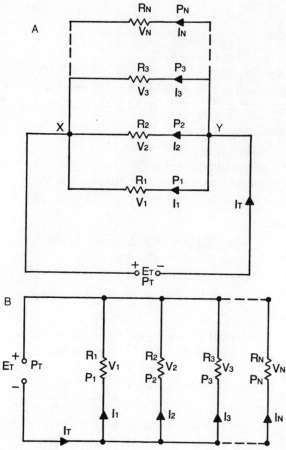

Fig. 3-12 Resistors in parallel.

where

$$I_T = \frac{V_1}{R_1} = \frac{E_T}{R_1}, I_2 = \frac{V_2}{R_2} = \frac{E_T}{R_2} \cdots$$

$$I_N = \frac{V_N}{R_N} = \frac{E_T}{R_N}$$

Note that the lowest value resistor carries the largest current. If the N resistors all have the same resistance, R,

$$I_T = NI, \quad \text{and} \quad I_1 = I_2 = \cdots = I_N = I = \frac{E_T}{R}$$

where I is the branch current through each resistor

3. The total equivalent conductance is equal to the sum of the individual conductances.

For N resistors in parallel,

$$G_T = G_1 + G_2 + G_3 + \cdots + G_N$$

$$\frac{1}{R_T} = \frac{1}{R_1} + \frac{1}{R_2} + \frac{1}{R_3} + \cdots + \frac{1}{R_N} \quad \text{(reciprocal formula)}$$

and

$$I_T = \frac{E_T}{R_T} = E_T G_T = E_T \times (G_1 + G_2 + G_3 + \cdots + G_N)$$

Note that the total equivalent resistance, R_T, is always less than that of the lowest value resistor in parallel.
For two resistors in parallel

$$R_T = \frac{R_1 \times R_2}{R_1 + R_2} \quad \text{(product-over-sum formula)}$$

If the N resistors all have the same resistance, $R = 1/G$,

$$G_T = NG, \ R_T = \frac{R}{N}$$

and

$$I_T = N \times \frac{E_T}{R}$$

4. The total power delivered from the source is equal to the total sum of the powers dissipated in the individual resistors.
For N resistors in parallel,

$$P_T = P_1 + P_2 + P_3 + \cdots + P_N$$

$$= I_T{}^2 R_T = \frac{E_T{}^2}{R_T} = E_T I_T$$

where

$$P_1 = \frac{E_T{}^2}{R_1} = I_1{}^2 R_1 = E_T I_1$$

etc.
Note that the lowest value resistor dissipates the most power. If the N resistors all have the same resistance, R,

$$P_T = NP \quad \text{and} \quad P_1 = P_2 = \cdots = P_N = P = I^2 R = \frac{E_T{}^2}{R} = E_T I$$

where I is the current in each branch

CURRENT-DIVISION RULE

5. The total current drawn from the source divides between the resistors in *inverse* proportion to the values of the resistors (current division rule). Then

$$I_1 = I_2 \times \frac{R_2}{R_1} = I_T \times \frac{R_T}{R_1} = I_T \times \frac{G_1}{G_T}, \text{ etc.}$$

For two resistors in parallel,

$$I_1 = I_T \times \frac{R_2}{R_1 + R_2}, \quad I_2 = I_T \times \frac{R_1}{R_1 + R_2}$$

The total power delivered from the source is also divided between the resistors in inverse proportion to the values of the resistors.

$$P_1 = P_2 \times \frac{R_2}{R_1} = P_T \times \frac{R_T}{R_1}, \text{ etc.}$$

THE SHORT CIRCUIT

6. The short circuit has theoretically zero resistance so that current can flow through the short circuit without developing any voltage drop. However, resistors themselves rarely become shorted, and the short circuit normally occurs as the result of bare connecting wires coming into contact or being joined together by a stray drop of solder. The short circuit is therefore a zero resistance path across, or in parallel with, the resistor.

Example 3-9

In Fig. 3-13, what is the value of the resistance between the points X and Y? What are the values of I_1, I_2, I_3, I_T, P_1, P_2, P_3, and P_T?

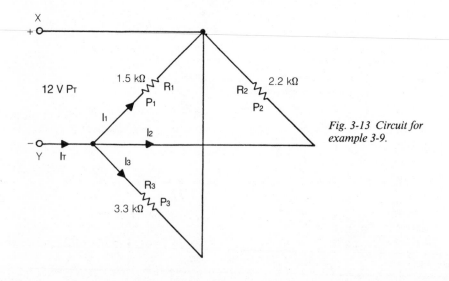

Fig. 3-13 Circuit for example 3-9.

Solution

All three resistors are connected effectively between X and Y and are therefore in parallel. To use the product-over-sum method for calculating the equivalent resistance, R_T, of three resistors in parallel, first obtain the equivalent resistance, R', of R_1 and R_2 in parallel. Then calculate R_T by finding the total resistance of R' and R_3 in parallel.

$$R' = \frac{R_1 \times R_2}{R_1 + R_2} = \frac{1.5 \times 2.2}{1.5 + 2.2} = \frac{1.5 \times 2.2}{3.7} = 0.892 \text{ k}\Omega$$

The total resistance between points X and Y is

$$R_T = \frac{R' \times R_3}{R' + R_3} = \frac{0.892 \times 3.3}{3.3 + 0.892} = \frac{0.892 \times 3.3}{4.192} = 0.702 \text{ k}\Omega = 702 \text{ }\Omega$$

Alternatively, the reciprocal formula may be used.

$$\frac{1}{R_T} = \frac{1}{1.5} + \frac{1}{2.2} + \frac{1}{3.3}$$
$$= 0.667 + 0.455 + 0.303 = 1.425$$
$$R_T = \frac{1}{1.425} = 0.702 \text{ k}\Omega = 702 \text{ }\Omega$$

The individual branch currents are

$$I_1 = \frac{E_T}{R_1} = \frac{12 \text{ V}}{1.5 \text{ k}\Omega} = 8.0 \text{ mA}$$

$$I_2 = \frac{E_T}{R_2} = \frac{12 \text{ V}}{2.2 \text{ k}\Omega} = 5.45 \text{ mA}$$

$$I_3 = \frac{E_T}{R_3} = \frac{12 \text{ V}}{3.3 \text{ k}\Omega} = 3.64 \text{ mA}$$

$$I_T = I_1 + I_2 + I_3 = 8.0 \text{ mA} + 5.45 \text{ mA} + 3.64 \text{ mA} = 17.09 \text{ mA}$$

Total resistance check. The total resistance is also given by

$$R_T = \frac{E_T}{I_T} = \frac{12 \text{ V}}{17.09 \text{ mA}} = 0.702 \text{ k}\Omega = 702 \text{ }\Omega$$

The total resistance therefore can be found either by repeated use of the product-over-sum formula or by the reciprocal method or by obtaining the total current from the branch currents (if a supply voltage is not given, any convenient value can be assumed).

$$P_1 = E_T I_1 = 12 \text{ V} \times 8.0 \text{ mA} = 96 \text{ mW}$$
$$P_2 = E_T I_2 = 12 \text{ V} \times 5.45 \text{ mA} = 65.4 \text{ mW}$$
$$P_3 = E_T I_3 = 12 \text{ V} \times 3.64 \text{ mA} = 43.7 \text{ mW}$$

Total power, $P_T = P_1 + P_2 + P_3 = 96 \text{ mW} + 65.4 \text{ mW} + 43.7 \text{ mW}$
$$= 205.1 \text{ mW}$$

Check: Total power delivered from source:

$$P_T = EI_T = 12 \text{ V} \times 17.09 \text{ mA} = 205.1 \text{ mW}$$

Example 3-10

In the circuit of Fig. 3-14, what are the values of I_1, P_1, P_2, and E_T?

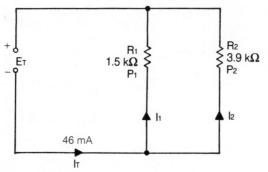

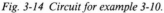

Fig. 3-14 Circuit for example 3-10.

Solution

$$I_1 = I_T \times \frac{R_2}{R_1 + R_2} = 46 \text{ mA} \times \frac{3.9 \text{ k}\Omega}{1.5 \text{ k}\Omega + 3.9 \text{ k}\Omega}$$

$$= 46 \times \frac{3.9}{5.4} = 33.2 \text{ mA}$$

$$P_1 = I_1^2 R_1 = (33.2 \text{ mA})^2 \times 1.5 \text{ k}\Omega = 1656 \text{ mW}$$

$$P_2 = P_1 \times \frac{R_1}{R_2} = 1656 \text{ mW} \times \frac{1.5 \text{ k}\Omega}{3.9 \text{ k}\Omega} = 637 \text{ mW}$$

Note that the lower value resistor, R_1, dissipates the greater power.

$$E_T = I_1 R_1 = 33.2 \text{ mA} \times 1.5 \text{ k}\Omega = 50 \text{ V} \quad \text{(rounded off)}$$

Example 3-11

A purely resistive load has a value of 3.7 kΩ that is reduced to 2.5 kΩ by connecting a single resistor across the load. What is the value of this resistor?

Solution

The resistor, R_X, is connected in parallel with the load, R_L, so that their combination has a total resistance of 2.5 kΩ.
This leads to

$$\frac{1}{R_T} = \frac{1}{R_L} + \frac{1}{R_X}$$

which leads to

$$R_X = \frac{R_L R_T}{R_L - R_T} = \frac{2.5 \times 3.7}{3.7 - 2.5}$$

$$= \frac{2.5 \times 3.7}{1.2} = 7.7 \ k\Omega$$

Example 3-12

Eight electric light bulbs, each with a rating of 5 W, 110 V, are being operated from a 110 V dc source. What is the value of the total load resistance?

Solution

The "hot" resistance of each bulb equals

$$\frac{E^2}{P} = \frac{110^2}{75} \ \Omega.$$

Because the bulbs must be connected in parallel, the total load resistance is

$$\frac{110^2/75}{8} = 20 \ \Omega \quad (\text{rounded off})$$

Example 3-13

A 20,000-ohm 200-watt resistor, a 40,000-ohm 100-watt resistor, and a 5000-ohm 50-watt resistor are connected in parallel. What is the maximum value of the total source current that will not cause the wattage rating of any of the resistors to be exceeded?

Solution

Because $E = \sqrt{PR}$, the maximum voltage ratings for the individual resistors are $\sqrt{20,000 \times 200}$, $\sqrt{40,000 \times 100}$, and $\sqrt{5,000 \times 50}$. The smallest of these ratings is $\sqrt{5000 \times 50} = \sqrt{250,000} = \sqrt{25 \times 10^4} = 10^2 \times \sqrt{25} = 500$ V.

$$\text{total current} = \frac{500}{20,000} + \frac{500}{40,000} + \frac{500}{5000}$$

$$= 0.025 + 0.0125 + 0.1$$

$$= 0.1374 \ A = 137.5 \ mA$$

Example 3-14

In Fig. 3-15, point E is grounded and a short appears across the 470 Ω resistor. What are the new potentials at the points A, B, C, and D?

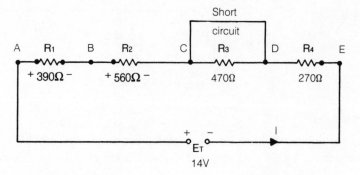

Fig. 3-15 Circuit for example 3-14.

Solution

Because of the short circuit, the resistance between the points C and D is zero. The new total equivalent resistance is

$$R_T = 390 + 560 + 270 = 1220 \ \Omega$$

$$\text{current, } I = \frac{E_T}{R_T} = \frac{14 \text{ V}}{1220 \ \Omega} = 0.0114754 \text{ A}$$

$$V_1 = IR_1 = 0.0114754 \text{ A} \times 390 \ \Omega = 4.475 \text{ V}$$
$$V_2 = IR_2 = 0.0114754 \text{ A} \times 560 \ \Omega = 6.426 \text{ V}$$
$$V_3 = 0 \text{ V}$$
$$V_4 = IR_4 = 0.0114754 \text{ A} \times 270 \ \Omega = 3.099 \text{ V}$$

potential at point A $= +14$ V
potential at point B $= +14 - 4.475 = +9.525$ V
potential at points C and D $= +9.525 - 6.426 = +3.099$ V

THE VOLTAGE-DIVIDER CIRCUIT

This circuit is a practical result of the voltage-division rule. By using it, a number of different voltages can be made available from a single voltage source.

In Fig. 3-16, V_1 is the voltage drop across the series combination of R_1, R_2, R_3, R_4, R_5 and is therefore equal to the source voltage, E_T. V_2 is dropped across R_2, R_3, R_4, R_5 in series:

$$V_2 = E_T \times \frac{(R_2 + R_3 + R_4 + R_5)}{R_T} = E_T \times \frac{(R_2 + R_3 + R_4 + R_5)}{(R_1 + R_2 + R_3 + R_4 + R_5)}$$

Similarly,

$$V_3 = E_T \times \frac{(R_3 + R_4 + R_5)}{R_T} = E_T \times \frac{(R_3 + R_4 + R_5)}{(R_1 + R_2 + R_3 + R_4 + R_5)}$$

$$V_4 = E_T \times \frac{(R_4 + R_5)}{R_T} = E_T \times \frac{(R_4 + R_5)}{(R_1 + R_2 + R_3 + R_4 + R_5)}$$

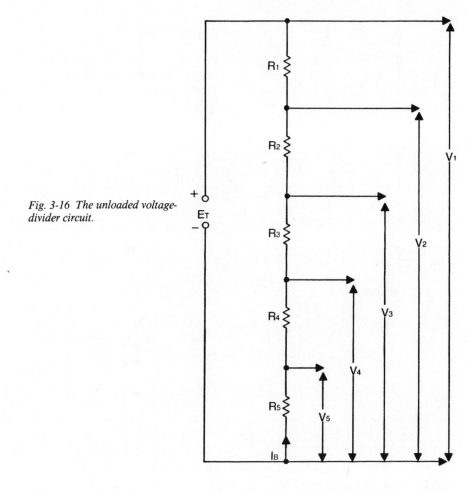

Fig. 3-16 The unloaded voltage-divider circuit.

and

$$V_5 = E_T \times \frac{R_5}{R_T} = E_T \times \frac{R_5}{(R_1 + R_2 + R_3 + R_4 + R_5)}$$

The voltages V_1, V_2, V_3, V_4, and V_5 develop because of the flow of the bleeder current, I_B, through the resistors. The availability of the different output voltages therefore is achieved at the expense of the power dissipated in the series string.

The voltage divider of Fig. 3-16 is operating under no-load conditions. As soon as loads are connected across the voltages V_1, V_2, V_3, V_4, and V_5, the circuit becomes a series-parallel arrangement and the four preceding equations are no longer valid.

Figure 3-17 illustrates a loaded voltage-divider circuit. The three load voltages are V_{L1}, V_{L2}, and V_{L3}, and these are associated respectively with load currents I_{L1}, I_{L2}, and I_{L3}. The bleeder current, I_B, is typically 10% of the total load current, $I_{LT} = I_{L1} + I_{L2} + I_{L3}$.

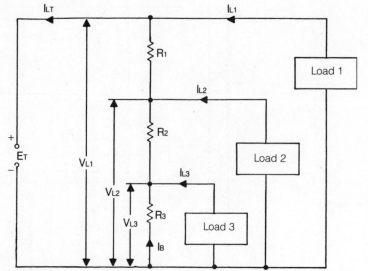

Fig. 3-17 The loaded voltage-divider circuit.

Then

$$V_{L1} = I_B R_3 \quad \text{and} \quad R_3 = \frac{V_{L3}}{I_B}$$

$$V_{L2} = V_{L3} + R_2 (I_{L3} + I_B)$$

and

$$R_2 = \frac{V_{L2} - V_{L3}}{I_{L3} + I_B}$$

$$E_T = V_{L1} = V_{L2} + R_1(I_{L2} + I_{L3} + I_B)$$

$$R_T = \frac{V_{L1} - V_{L2}}{I_{L2} + I_{L3} + I_B}$$

Note that I_{L1} is drawn directly from the source voltage and therefore does not appear in the equations.

Given the values of the load voltages with their currents and assuming a certain percentage for the bleeder current, the values of R_1, R_2, R_3 can be calculated.

Example 3-15

In Fig. 3-18, find the values of V_1, V_2, V_3, V_4, and V_5.

Solution

$$V_1 = 75 \text{ V}$$

$$V_2 = 75 \times \frac{(100 + 20 + 4 + 1)}{(500 + 100 + 20 + 4 + 1)} = 75 \times \frac{125}{625} = 15 \text{ V}$$

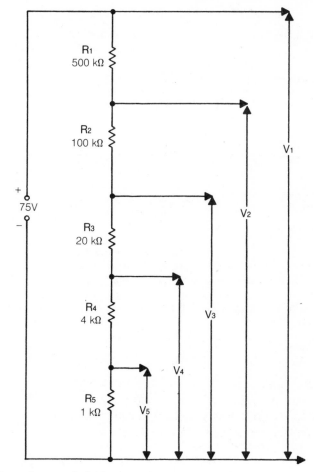

Fig. 3-18 Circuit for example 3-15.

$$V_3 = 75 \times \frac{(20 + 4 + 1)}{(500 + 100 + 20 + 4 + 1)} = 75 \times \frac{25}{625} = 3 \text{ V}$$

$$V_4 = 75 \times \frac{(4 + 1)}{(500 + 100 + 20 + 4 + 1)} = 75 \times \frac{5}{625} = 0.6 \text{ V}$$

$$V_5 = 75 \times \frac{(1)}{(500 + 100 + 20 + 4 + 1)} = 75 \times \frac{1}{625} = 0.12 \text{ V}$$

Fig. 3-18 could be referred to as a "divide-by-5" circuit because V_2 is one-fifth of V_1, V_3 is 1/5th of V_2, etc.

Example 3-16

Referring to Fig. 3-17, $V_{L1} = E_T = 24$ V, $V_{L2} = 15$ V, $V_{L3} = 9$ V, $I_{L3} = 50$ mA, $I_{L2} = 85$ mA, and $I_{L1} = 125$ mA. If the bleeder current is 10% of the total load current, what are the values required for R_1, R_2, and R_3?

Solution

total load current, $I_{LT} = I_{L1} + I_{L2} + I_{L3} = 125 + 85 + 50$
$$= 260 \text{ mA}$$

bleeder current, $I_B = \dfrac{10}{100} \times 260 = 26 \text{ mA}$

$$R_3 = \frac{V_{L3}}{I_B} = \frac{9 \text{ V}}{26 \text{ mA}} = 346 \ \Omega$$

$$R_2 = \frac{V_{L2} - V_{L3}}{I_{L3} + I_B} = \frac{15 \text{ V} - 9 \text{ V}}{50 \text{ mA} + 26 \text{ mA}} = \frac{6 \text{ V}}{76 \text{ mA}} = 79 \ \Omega$$

$$R_1 = \frac{V_{L1} - V_{L2}}{I_{L2} + I_{L3} + I_B} = \frac{24 \text{ V} - 15 \text{ V}}{85 \text{ mA} + 50 \text{ mA} + 26 \text{ mA}} = \frac{9 \text{ V}}{161 \text{ mA}} = 56 \ \Omega$$

The power dissipated in these resistors is

$$P_1 = 9 \text{ V} \times 161 \text{ mA} = 1449 \text{ mW}$$
$$P_2 = 6 \text{ V} \times 76 \text{ mA} = 456 \text{ mW}$$
$$P_3 = 9 \text{ V} \times 26 \text{ mA} = 234 \text{ mW}$$

Suitable resistors would be R_1: 56 Ω, 2 W, R_2: 82 Ω, 1 W, and R_3: 330 Ω, 1/2 W.

SERIES-PARALLEL RESISTOR NETWORKS

A series-parallel circuit can be defined as one in which parts of the circuit have the properties of a simple series arrangement while other parts have the properties of a basic parallel connection. There is an infinite variety of such circuits, but one example appears in Fig. 3-19A. Its analysis can be carried out in a number of steps.

Step 1.   Identify all single resistors that are directly in series with each other, or in other words, those resistors that carry the same current. Combine all such series resistors into their equivalent resistances. In the example chosen, R_2 and R_3 are in series and their total resistance is 11 kΩ + 4 kΩ = 15 kΩ. As far as the equivalent resistance is concerned, the order in which the components are connected is unimportant; therefore R_4 and R_5 are also in series with a total resistance of 1 kΩ + 3 kΩ = 4 kΩ. The circuit now can be redrawn as shown in Fig. 3-19B.

Step 2.   Identify all single resistors that are directly in parallel with each other. These resistors will be connected between points in which a current division occurs so that R_6, R_7, and R_8 are in parallel and their equivalent resistance, R_{eq}, is given by

$$\frac{1}{R_{eq}} = \frac{1}{6} + \frac{1}{3} + \frac{1}{2} = \frac{1}{1}$$
$$R_{eq} = 1 \text{ k}\Omega$$

The circuit now can be drawn again as in Fig. 3-19C.

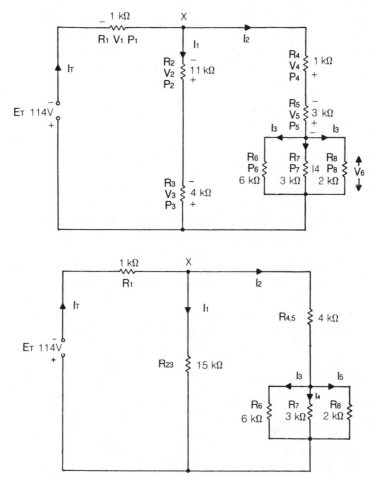

Fig. 3-19 A&B Combining the series resistors.

Step 3. Combine all equivalent resistances that are in series. This applies to the equivalent resistances of R_4, R_5, R_6, and R_7, so that the total resistance of these five resistors is 4 kΩ + 1 kΩ = 5 kΩ. Again the circuit is redrawn as in Fig. 3-19D.

Step 4. Combine all equivalent resistances that are in parallel. In Fig. 3-19D, a current split occurs at point X so that the equivalent resistances of R_2, R_3, and R_4, R_5, R_6, R_7, and R_8 are in parallel. The total resistance of these seven resistors is

$$\frac{5 \times 15}{5 + 15} = \frac{5 \times 15}{20} = 3.75 \text{ k}\Omega$$

The final redrawing of the circuit is then shown in Fig. 3-19E.

Step 5. The total resistance, R_T, of the whole circuit is the series combination of R_1 and the equivalent resistance of R_2, R_3, R_4, R_5, R_6, R_7, and R_8. R_T is therefore equal to 1 kΩ + 3.75 kΩ = 4.75 kΩ.

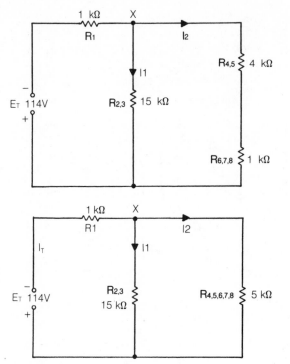

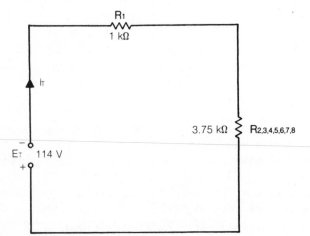

Fig. 3-19 C&D Combining the parallel resistors.

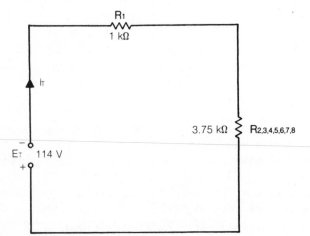

Fig. 3-19 E. Simplified circuit showing the equivalent resistances.

In carrying out the steps to find R_T, the equivalent resistances of series and parallel combinations were calculated alternately. The circuit was broken down a bit at a time with each redrawing being simpler than the one before. After some experience with series-parallel circuits, the intermediate drawings can be omitted. The strategy with these problems is to start furthest away from the source and gradually work towards the source.

The next stage in the analysis is to work away from the source and determine the individual currents, voltage drops and powers dissipated in the various resistors. When calculating the required values it is important to check that the voltages around any closed loop exactly balance (Kirchhoff's Voltage Law). At the same time the currents existing at any point must also balance (Kirchhoff's Current Law).

$$\text{total current, } I_T = \frac{E_T}{R_T} = \frac{114 \text{ V}}{3.75 \text{ k}\Omega} = 24 \text{ mA}$$

$$V_1 = I_T R_1 = 24 \text{ mA} \times 1 \text{ k}\Omega = 24 \text{ V}$$

By Kirchhoff's Voltage Law,

$$V_2 + V_3 = E_T - V_1 = 114 \text{ V} - 24 \text{ V} = 90 \text{ V}$$

Then,

$$I_1 = \frac{V_2 + V_3}{R_2 + R_3} = \frac{90 \text{ V}}{11 \text{ k}\Omega + 4 \text{ k}\Omega} = 6 \text{ mA}$$

$$V_2 = I_1 R_2 = 6 \text{ mA} \times 11 \text{ k}\Omega = 66 \text{ V}$$
$$V_3 = I_1 R_3 = 6 \text{ mA} \times 4 \text{ k}\Omega = 24 \text{ V}$$

By Kirchhoff's Current Law:

$$I_2 = I_T - I_1 = 24 \text{ mA} - 6 \text{ mA} = 18 \text{ mA}$$

Then

$$V_4 = I_2 R_4 = 18 \text{ mA} \times 1 \text{ k}\Omega = 18 \text{ V}$$
$$V_5 = I_2 R_5 = 18 \text{ mA} \times 3 \text{ k}\Omega = 54 \text{ V}$$

By Kirchhoff's Voltage Law,

$$V_6 = 90 \text{ V} - V_4 - V_5 = 90 \text{ V} - 18 \text{ V} - 54 \text{ V} = 18 \text{ V}$$

Then

$$I_3 = \frac{V_6}{R_6} = \frac{18 \text{ V}}{6 \text{ k}\Omega} = 3 \text{ mA}$$

$$I_4 = \frac{V_6}{R_7} = \frac{18 \text{ V}}{3 \text{ k}\Omega} = 6 \text{ mA}$$

$$I_5 = \frac{V_6}{R_8} = \frac{18 \text{ V}}{2 \text{ k}\Omega} = 9 \text{ mA}$$

$$P_1 = I_T V_1 = 24 \text{ mA} \times 24 \text{ V} = 576 \text{ mW}$$
$$P_2 = I_1 V_2 = 6 \text{ mA} \times 66 \text{ V} = 396 \text{ mW}$$
$$P_3 = I_1 V_3 = 6 \text{ mA} \times 24 \text{ V} = 144 \text{ mW}$$
$$P_4 = I_2 V_4 = 18 \text{ mA} \times 18 \text{ V} = 324 \text{ mW}$$
$$P_5 = I_2 V_5 = 18 \text{ mA} \times 54 \text{ V} = 972 \text{ mW}$$
$$P_6 = I_3 V_6 = 3 \text{ mA} \times 18 \text{ V} = 54 \text{ mW}$$
$$P_7 = I_4 V_6 = 6 \text{ mA} \times 18 \text{ V} = 108 \text{ mW}$$
$$P_8 = I_5 V_6 = 9 \text{ mA} \times 18 \text{ V} = 162 \text{ mW}$$
$$\text{total power dissipated: } P_T = 2736 \text{ mW}$$

Check: Total power delivered from the source:

$$P_T = I_T E_T = 24 \text{ mA} \times 114 \text{ V} = 2736 \text{ mW}$$

There are many complex resistor networks that cannot be treated as simple series-parallel combinations. Figure 3-20 shows one example in which the position of R_5 in relation to the other resistors cannot be specified as either series or parallel. To analyze such a circuit requires some of the more sophisticated techniques that I discuss in Chapter 8.

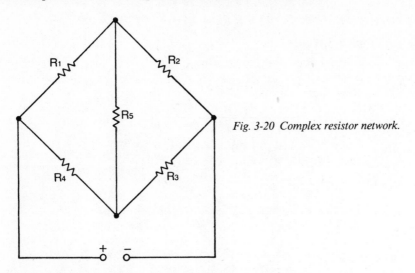

Fig. 3-20 *Complex resistor network.*

CHAPTER SUMMARY

☐ Resistors in Series

$$E_T = V_1 + V_2 + V_3 + \cdots + V_N \quad \text{(Kirchhoff's Voltage Law)}$$
$$R_T = R_1 + R_2 + R_3 + \cdots + R_N$$
$$P_T = P_1 + P_2 + P_3 + \cdots + P_N$$

$$V_1 = V_2 \times \frac{R_1}{R_2} = E_T \times \frac{R_1}{R_T}$$

$$P_1 = P_2 \times \frac{R_1}{R_2} = P_T \times \frac{R_1}{R_T}$$

☐ Series-Dropping Resistor

$$V_D = E - V_L = E - I_L R_L$$

$$R_D = \frac{V_D}{I_L}$$

$$P_D = I_L \times V_D$$

☐ Resistors in Parallel

$$I_T = I_1 + I_2 + I_3 + \cdots + I_N \quad \text{(Kirchhoff's Current Law)}$$

$$\frac{1}{R_T} = \frac{1}{R_1} + \frac{1}{R_2} + \frac{1}{R_3} + \cdots + \frac{1}{R_N}$$

$$P_T = P_1 + P_2 + P_3 + \cdots + P_N$$

$$I_1 = I_2 \times \frac{R_2}{R_1} = \frac{I_T \times R_T}{R_1}$$

$$P_1 = P_2 \times \frac{R_2}{R_1} = P_T \times \frac{R_T}{R_1}$$

For two resistors in parallel,

$$R_T = \frac{R_1 R_2}{R_1 + R_2}$$

$$I_1 = I_T \times \frac{R_2}{R_1 + R_2}, I_2 = I_T \times \frac{R_1}{R_1 + R_2}$$

4

Voltage
and Current Sources

IN THE STUDY of actual electrical sources, you will learn:

□ How you can represent an electrical source by a constant-voltage model.

□ How to develop maximum power in the load by the process of matching.

□ That you can obtain a higher voltage by connecting a number of practical sources in series.

□ How you can represent an electrical source by a constant-current model.

□ How to reduce the equivalent internal resistance by connecting a number of practical sources in parallel.

□ How to obtain a higher voltage and also reduce the equivalent internal resistance by connecting a number of practical sources in a series-parallel arrangement.

Most practical electrical sources such as a battery generate a terminal potential difference, or PD, that falls as the load current is increased. This effect can be explained by assuming that the source contains a constant-voltage EMF, E, which is independent of the load current, in series with an internal resistance, R_i. This equivalent circuit is shown in Fig. 4-1. If the load resistance, R_L, is reduced, the load current, I_L, will increase. This raises the voltage drop, V_i, across the internal resistance so that the terminal PD or load voltage, V_L, falls.

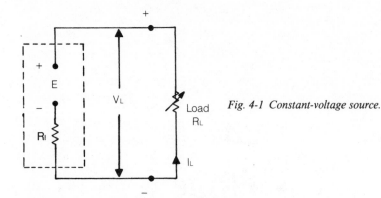

Fig. 4-1 Constant-voltage source.

CONSTANT-VOLTAGE SOURCES

You can obtain the value of E by removing the load and then connecting a voltmeter across the terminals. The voltmeter is assumed to draw negligible current from the source so that the voltage drop across the internal resistance is zero and the terminal PD is equal to the constant-voltage EMF. The value of E therefore is referred to as the *open-circuit terminal voltage.*

If you connect a conductor of negligible resistance between the terminals, R_L is zero and the current will be limited only by the internal resistance. Because the terminals have been short-circuited effectively, the large current that flows is called the *short-circuit current* whose value is E/R_i. The internal resistance is therefore given by

$$\text{internal resistance, } R_i = \frac{\text{open-circuit terminal voltage, E}}{\text{short-circuit terminal current, E/}R_i}$$

Normally the short-circuit current is so large that the voltage source will be damaged, if not destroyed. A more practical method for finding the internal resistance is to adjust the value of R_i until $V_L = E/2$; R_L and R_i will then be equal.

The equations related to Fig. 4-1 are

$$\text{load current, } I_L = \frac{V_L}{R_L} = \frac{V_i}{R_i} = \frac{E}{R_L + R_i}$$

$$\text{load voltage, } V_L = I_L R_L = \frac{E \times R_L}{R_L + R_i} = E - V_i = E - I_L R_i$$

$$\text{internal voltage drop} = I_L R_i = E - V_L.$$

As an example, consider an electrical source with an open circuit terminal voltage of 12 V and an internal resistance of 0.15 Ω. The load voltage, V_L, is calculated for various load currents (see Table 4-1).

Table 4-1. Load Voltages and Load Currents

I_L Amperes	V_i Volts	V_L Volts	R_L Ohms
0	0	12	Open circuit. Infinite ohms
10	1.5	10.5	1.05
20	3	9	0.45
30	4.5	7.5	0.25
40	6	6	0.15
50	7.5	4.5	0.09
60	9	3	0.05
80	12	0	Short circuit. Zero ohms

The graph of load voltage/load current appears in Fig. 4-2. The graph is a straight line because the internal resistance is constant and can be calculated from the line's negative slope.

$$R_i = \frac{\Delta V_L}{\Delta I_L} = \frac{1.5 \text{ V}}{10 \text{ A}} = 0.5 \text{ }\Omega$$

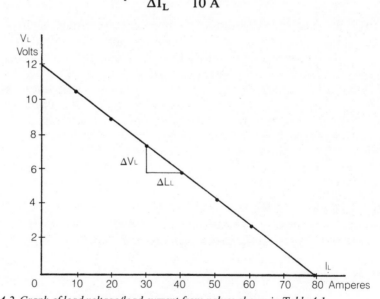

Fig. 4-2 Graph of load voltage/load current from values shown in Table 4-1.

In many cases, the internal resistance of a power source is not constant, and then the V_L/I_L graph will be a regulation curve. The term "voltage regulation" is a measure of the extent to which V_L changes as I_L varies. Voltage regulation is expressed as a percentage and is calculated from

$$\text{percentage regulation} = \frac{V_{NL} - V_{FL}}{V_{FL}} \times 100\%$$

where

V_{NL} = terminal voltage under no-load conditions
V_{FL} = terminal voltage under full-load conditions

Assuming that the open circuit voltage is the no-load condition, $V_{NL} =$ E and $V_{FL} = ER_{FL}/(R_{FL} + R_i)$. Then

$$\text{percentage regulation} = \frac{\left(E - \dfrac{ER_{FL}}{R_{FL} + R_i}\right)}{\dfrac{ER_{FL}}{R_{FL} + R_i}} \times 100 = \frac{R_i}{R_{FL}} \times 100\%$$

where

R_{FL} = load resistance corresponding to full-load conditions

In Table 4-1, V_{NL} is 12 V. If the full load current is 20 A, V_{FL} is 9 V. Then

$$\text{percentage regulation} = \frac{12\ V - 9\ V}{9\ V} \times 100$$

$$= 33\tfrac{1}{3}\%$$

An ideal voltage source would have zero percentage regulation, in which case the regulation "curve" would be a horizontal line. This would indicate that the load voltage was constant and independent of the load current. Therefore, as seen from Table 4-1, the condition for a high-load voltage requires that the load resistance be large compared to the internal resistance; typically R_L must be at least five to ten times the value of R_i. The constant voltage EMF then divides between the internal and load resistances and most of the voltage appears across the load. By the voltage division rule, $V_L/V_i = R_L/R_i$ and $V_L/E = R_L/(R_L + R_i)$.

Example 4-1

In Fig. 4-1, E = 42 V and $R_i = 0.1\ \Omega$. Under no-load conditions R_L is an open circuit while the full-load value of R_L is 1.5 Ω. Calculate the full-load values of I_L, V_L, and the percentage of regulation.

Solution

$$\text{full-load voltage, } V_{FL} = 42\ V \times \frac{1.5\ \Omega}{1.5\ \Omega + 0.1\ \Omega} = 39.375\ V$$

$$\text{full-load current, } I_L = \frac{V_{FL}}{R_L} = \frac{39.375\ V}{1.5\ \Omega} = 26.25\ A$$

$$\text{no-load voltage, } V_{NL} = 42.0\ V$$

$$\text{percentage of regulation} = \frac{42.0 - 39.375}{39.375} \times 100 = 6.67\%$$

Check:

$$\text{percentage regulation} = \frac{R_i}{R_{FL}} \times 100 = \frac{0.1}{1.5} \times 100 = 6.67\%$$

CONDITION FOR MAXIMUM POWER TRANSFER

The power in the load, P_L, is equal to the product of V_L and I_L and is therefore zero under both the open-circuit and the short-circuit conditions. Between these extremes, the load power reaches a maximum value corresponding to a particular value of R_L. There are many examples in electronics in which it is necessary to know the condition for maximum power transfer to the load. The power developed in the load is

$$P_L = I_L{}^2 R_L = \left(\frac{E}{(R_L + R_i)}\right)^2 \times R_L$$

At this point, it is customary to use calculus to derive the condition for P_L to reach its maximum value as R_L and hence I_L are varied. However, it is possible to obtain the condition by algebra alone. If P_L is to reach its maximum value, its reciprocal $1/P_L$ must fall at the same time to its minimum value. Then

$$\frac{1}{P_L} = \frac{(R_L + R_1)^2}{E^2 R_L} = \frac{1}{E^2}\left[\frac{(R_L - R_L)^2}{R_L} + 4\,R_i\right]$$

The minimum value of a square such as $(R_L - R_i)^2$ is zero; this occurs when

$$R_L = R_i.$$

The condition for maximum power transfer therefore requires that the load resistance, R_L, is *matched* (made equal) to the internal resistance, R_i. For example, a receiver's loadspeaker is matched to the transistor used in the final stage; this enables the audio power developed in this stage to be transferred to the load of the loudspeaker.

When R_L and R_i are matched, the maximum power developed in the load is

$$P_{L\,max} = \frac{E^2}{4\,R_i}$$

Another factor that must be taken into account is the *circuit efficiency*. This is defined as the percentage of the power developed in the load as compared to the total power drawn from the constant voltage EMF. Therefore

$$\text{circuit efficiency} = \frac{I_L{}^2 R_L}{I_L{}^2(R_L + R_i)} \times 100\% = \frac{R_L}{R_L + R_i} \times 100\%$$

Because $V_L/E = R_L/(R_L + R_i)$, the condition for a large voltage across the load is the same as for a high-circuit efficiency.

These results are best illustrated by the example illustrated in Fig. 4-3. The constant voltage EMF is 24 V, and this is associated with an internal resistance of 4 Ω. The load resistance, R_L, is varied from zero (short circuit) to infinity (open circuit) and for certain values of R_L, the values of I_L, V_L, P_L, and percentage efficiency are calculated as shown in Table 4-2. Figure 4-4 contains the graphs of V_L, I_L, P_L, and percentage of efficiency plotted against R_L. Another presentation is in Fig. 4-5, which shows the relationships of I_L versus V_L, P_L, and percentage of efficiency. The V_L/I_L and percentage of efficiency/I_L graphs are then straight lines while the P_L/I_L curve is a parabola.

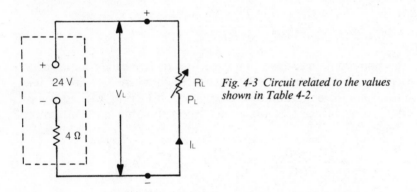

Fig. 4-3 Circuit related to the values shown in Table 4-2.

Table 4-2. How Load Resistance Affects Efficiency

R_L Ohms	I_L Amperes	V_L Volts	P_L Watts	% Efficiency
Open circuit. Infinite Ohms	0	24	0	100
20	1	20	20	83.3
12	1.5	18	27	75
8	2	16	32	66.7
4	3	12	36	50
2	4	8	32	33.3
Short circuit. Zero Ohms.	6	0	0	0

Example 4-2

In Fig. 4-1, E = 36 V, R_i = 2.5 Ω, and R_L = 8 Ω. What additional load must be added in parallel with R_L in order for maximum power to be developed in the new total load? Calculate the values of P_L before and after the introduction of the additional load.

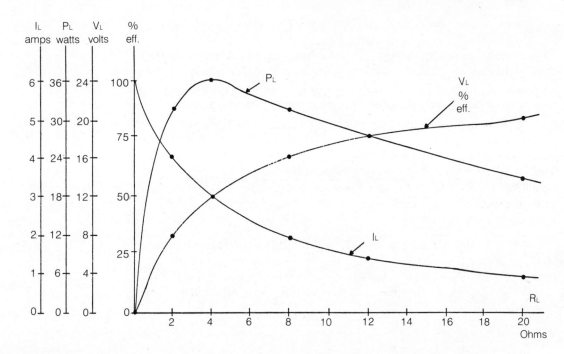

Fig. 4-4 Graphs of load voltage, load current, load power, and percentage of efficiency against load resistance.

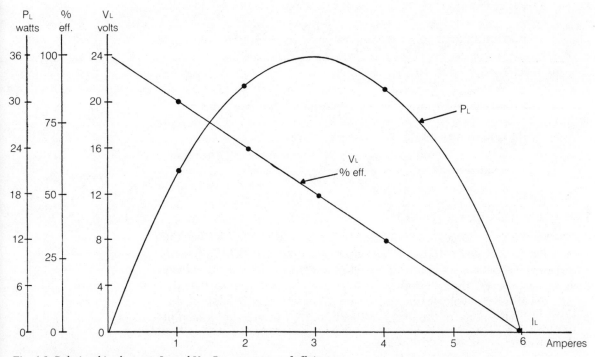

Fig. 4-5 Relationships between IL and VL, PL, percentage of efficiency.

Solution

For maximum power transfer, the total load must be matched to the internal resistance of 2.5 Ω.

$$\text{value of the additional load} = \frac{2.5 \times 8}{8 - 2.5} = \frac{20}{5.5} = 3.64 \ \Omega$$

$$\text{original load power} = \frac{E^2 R_L}{(R_i + R_L)^2} = \frac{(36 \text{ V})^2 \times 8 \ \Omega}{(2.5 \ \Omega + 8 \ \Omega)^2} = 94.0 \text{ W}$$

$$\text{final (maximum) load power} = \frac{E^2}{4 \ R_i} = \frac{(36 \text{ V})^2}{4 \times 2.5 \ \Omega} = 129.6 \text{ W}$$

Example 4-3

A voltage source has a constant EMF of 28 volts and an internal resistance of 3 Ω. It is connected to a load, R_L, which is in series with an additional resistance of 5 Ω. What value of R_L will allow maximum power transfer to the load? Calculate the load power if this value of R_L is (a) doubled and (b) halved.

Solution

For maximum power transfer the value of R_L must be matched to the total of all the resistances not associated with the load. Therefore R_L must equal 5 Ω + 3 Ω = 8 Ω.

$$\text{maximum load power} = \frac{E^2}{4 \ R_L} = \frac{(26 \text{ V})^2}{4 \times 8 \ \Omega} = 24.5 \text{ W}$$

(a) If R_L is doubled to 16 Ω,

$$\text{load power} = \frac{(28 \text{ V})^2 \times 16 \ \Omega}{(3 \ \Omega + 5 \ \Omega + 16 \ \Omega)^2} = 21.8 \text{ W}$$

(b) If R_L is halved to 4 Ω,

$$\text{load power} = \frac{(28 \text{ V})^2 \times 4 \ \Omega}{(3 \ \Omega + 5 \ \Omega + 4 \ \Omega)^2} = 21.8 \text{ W}$$

VOLTAGE SOURCES IN SERIES

When voltage sources are connected in series-aiding as in Fig. 4-6, the total constant voltage EMF, E_T, is the sum of the individual EMFs. Therefore the normal purpose of connecting cells in series is to increase the voltage available. However, the internal resistances are also additive so that the greater the number of voltage sources, the worse is the voltage regulation; this is the real disadvantage of the series connection. If N sources are joined in series-aiding,

$$E_T = E_1 + E_2 + E_3 + \cdots + E_N$$

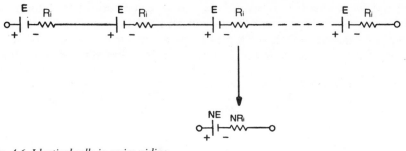

Fig. 4-6 Identical cells in series-aiding.

and

$$R_{iT} = R_{i1} + R_{i2} + R_{i3} + \cdots + R_{iN}.$$

If each source has a constant voltage EMF, E, and an internal resistance R_i,

$$E_T = NE$$

and

$$R_{iT} = NR_i$$

Example 4-4

Figure 4-6 represents eight voltage sources, each with an EMF of 1.5 V and an internal resistance of 0.1 Ω. What load resistance can be connected across the sources for maximum power transfer? Calculate the value of the maximum load power.

Solution

$$\text{total EMF, } E_T = 8 \times 1.5 \text{ V} = 12 \text{ V}$$
$$\text{total internal resistance, } R_{iT} = 8 \times 0.1 \ \Omega = 0.8 \ \Omega$$
$$\text{for maximum power transfer, } R_L = R_{iT} = 0.8 \ \Omega$$
$$\text{maximum load power} = \frac{E_T{}^2}{4 \, R_L} = \frac{(12 \text{ V})^2}{4 \times 0.8 \ \Omega} = 45 \text{ W}$$

CONSTANT-CURRENT SOURCES

The fact that the load voltage falls as the load current increases was explained by assuming that the source contained a constant voltage EMF in series with an internal resistance. However, this is only one possible assumption. The alternative is to consider that the source is a constant-current generator that produces a variable EMF and has the same value of internal resistance in parallel. The generator current is equal to the short-circuit value as already discussed.

The equivalence of the constant-voltage and constant-current generators is best illustrated by an example in which the open circuit voltage is 24 V and the short circuit current is 6 A. The internal resistance is therefore 24 V/6 A = 4 Ω. The circuits for the constant-voltage and constant-current generators appear in Figs. 4-7A and B. As a convention, the arrow in the symbol for the current generator indicates the direction of the electron flow.

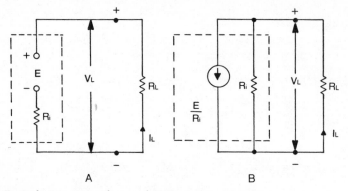

A   B

Fig. 4-7 Equivalent constant-voltage and constant-current generators.

Let a load of 8 Ω be connected across each of the generators. In the constant voltage circuit,

$$V_L = 24 \text{ V} \times \frac{8 \text{ }\Omega}{8 \text{ }\Omega + 4 \text{ }\Omega} = 16 \text{ V}$$

and therefore

$$I_L = 16 \text{ V}/8 \text{ }\Omega = 2 \text{ A}.$$

For the constant-current generator,

$$I_L = 6 \text{ A} \times \frac{4 \text{ }\Omega}{8 \text{ }\Omega + 4 \text{ }\Omega} = 2 \text{ A}$$

and

$$V_L = 2 \text{ A} \times 8 \text{ }\Omega = 16 \text{ V}.$$

Therefore, as far as the external load is concerned, the concepts of constant-voltage and constant-current generators are equally valid. However, the generator circuits themselves are not exact equivalents because under open-circuit load conditions, power is dissipated in the internal resistance of the current generator but not in the internal resistance of the voltage generator.

The equations for the generators are:

1. **Constant voltage generator**   Constant voltage EMF = open circuit terminal voltage.

$$\text{internal resistance} = \frac{\text{open circuit terminal voltage}}{\text{short circuit terminal current}}$$

2. **Constant current generator** Constant current = short-circuit terminal current.

$$\text{internal resistance} = \frac{\text{open circuit terminal voltage}}{\text{short circuit terminal current}}$$

Example 4-5

A dc source has an open circuit terminal voltage of 100 V and a short-circuit current of 20 A. A 20 Ω load is connected across the terminals. Derive the values in the constant voltage and constant-current equivalent circuits, and calculate the load current in each case. What are the values of V_L, P_L and the percentage efficiency?

Solution

Constant voltage generator Constant voltage EMF = 100 V. Internal resistance = 100 V/20 A = 5 Ω.

$$\text{load voltage, } V_L = \frac{100 \text{ V} \times 20 \ \Omega}{20 \ \Omega + 5 \ \Omega} = 80 \text{ V}$$

$$\text{load current, } I_L = \frac{80 \text{ V}}{20 \ \Omega} = 4 \text{ A}$$

$$\text{load power, } P_L = 80 \text{ V} \times 4 \text{ A} = 320 \text{ W}$$
$$\text{total power, } P_T = 100 \text{ V} \times 4 \text{ A} = 400 \text{ W.}$$

$$\text{percentage efficiency} = \frac{P_L}{P_T} \times 100\% = \frac{320}{400} \times 100$$
$$= 80\%$$

Constant current generator Constant source current = 20 A

$$\text{load current, } I_L = 20 \text{ A} \times \frac{5 \ \Omega}{20 \ \Omega + 5 \ \Omega} = 4 \text{ A}$$

CURRENT SOURCES IN PARALLEL

When combining two or more sources in parallel, it is inconvenient to use constant-voltage equivalent circuits as shown in Fig. 4-8A. The main difficulty is the circulating currents that exist between the sources, so that at the very least, the analysis requires the use of Kirchhoff's Laws. However, if you replace each source with a constant-current generator (Fig. 4-8B), you can simply add the currents to produce the total generator current. However, if some of the sources are connected in some parallel-opposing fashion with respect to the load, you must sum the currents algebraically. The total equivalent internal resistance is obtained by applying the reciprocal formula to the separate parallel resistances. This method of analyzing parallel sources is an

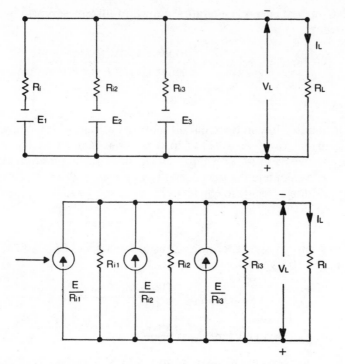

Fig. 4-8 Constant-voltage and constant-current equivalent circuits for cells in parallel.

application of Millman's Theorem (Chapter 8). If required, you can transform the final current generator into its constant-voltage equivalent.

If N sources are connected in parallel-aiding (all terminals of the same polarity are joined together), then

$$I_T = I_1 + I_2 + \cdots + I_N$$

$$R_{iT} = \cfrac{1}{\cfrac{1}{R_{i1}} + \cfrac{1}{R_{i2}} + \cfrac{1}{R_{i3}} + \cdots + \cfrac{1}{R_{iN}}}$$

If the N parallel sources are all identical,

$$I_T = NI$$

$$R_{iT} = \frac{R_i}{N}$$

where I and R_i are the constant current and equivalent resistance of one source. Total equivalent voltage, $E_T = I_T \times R_{iT} = NI \times R_i/N = I \times R_i =$ constant voltage EMF of one source (Fig. 4-9).

The advantage of connecting cells in parallel therefore is not to increase the voltage but to reduce the internal resistance. Normally equivalent constant-voltage generators are used for sources in series while parallel sources are replaced by their equivalent constant-current generators.

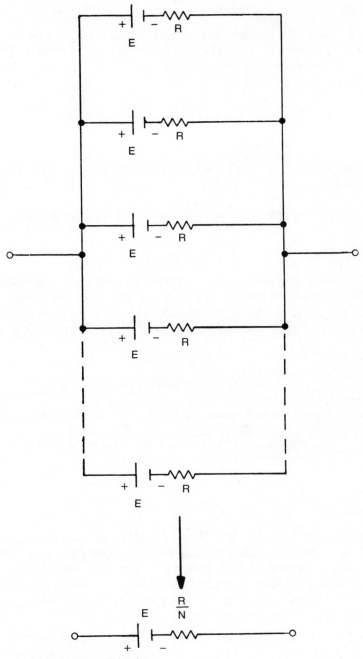

Fig. 4-9 Identical cells in parallel-aiding.

Example 4-6

In Fig. 4-8A, the values are $E_1 = 10$ V, $E_2 = 12$ V, $E_3 = 8$ V, $R_{i1} = 0.5$ Ω. $R_{i2} = 0.2$ Ω, $R_{i3} = 0.1$ Ω and $R_L = 2$ Ω. Calculate the value of V_L.

Solution

The values of the equivalent constant-current generators are $I_1 = 10$ V/ 0.5 $\Omega = 20$ A, $I_2 = 12$ V/0.2 $\Omega = 60$ A, $I_3 = 8$ V/0.1 $\Omega = 80$ A. The total equivalent current $= 20$ A $+ 60$ A $+ 80$ A $= 160$ A. The total internal resistance is given by

$$R_{iT} = \frac{1}{\dfrac{1}{0.5} + \dfrac{1}{0.2} + \dfrac{1}{0.1}} = \frac{1}{2 + 5 + 10} = \frac{1}{17} = 0.0588 \ \Omega$$

Constant voltage EMF $= 0.0588 \ \Omega \times 160$ A $= 9.408$ V. By the voltage division rule,

$$V_L = 9.408 \text{ V} \times \frac{2.0 \ \Omega}{2.0 \ \Omega + 0.0588 \ \Omega} = 9.1 \text{ V (rounded off)}$$

Example 4-7

Figure 4-9 represents eight sources that are connected in parallel aiding fashion. Each source has an EMF of 1.5 V and an internal resistance of 0.1 Ω. What load resistance can you connect across the sources for maximum power transfer? Calculate the value of the maximum load power.

Solution

$$\text{total EMF, } E_T = \text{EMF of one source} = 1.5 \text{ V}$$

$$\text{total internal resistance, } R_{iT} = \frac{0.1 \ \Omega}{8} = 0.0125 \ \Omega$$

$$\text{for maximum power transfer, } R_L = R_{it} = 0.0125 \ \Omega$$

$$\text{maximum load power} = \frac{(1.5 \text{ V})^2}{4 \times 0.0125 \ \Omega} = 45 \text{ W}$$

Example 4-8

Eight cells, each with an EMF of 1.5 V and an internal resistance of 0.1 Ω are connected in series-aiding fashion. Eight such banks are then connected in parallel-aiding fashion for a total of 64 cells. With such an arrangement, what is the total EMF and the total internal resistance? What value of load resistance will provide maximum power transfer and what is the value of the maximum load power?

Solution

$$\text{total EMF, } E_T = 8 \times 1.5 \text{ V} = 12 \text{ V}$$

$$\text{total internal resistance, } R_{iT} = \frac{8 \times 0.1 \ \Omega}{8} = 0.1 \ \Omega$$

$$\text{for maximum power transfer, } R_L = R_{iT} = 0.1 \ \Omega$$

$$\text{maximum load power} = \frac{(12 \text{ V})^2}{4 \times 0.1 \ \Omega} = 360 \text{ W}$$

CHAPTER SUMMARY

☐ Constant Voltage Source

constant voltage EMF, E = open-circuit terminal voltage

$$\text{internal resistance, } R_i = \frac{\text{open-circuit terminal voltage, E}}{\text{short-circuit terminal current, E/R}_i}$$

$$\text{load current, } I_L = \frac{V_L}{R_L} = \frac{V_i}{R_i} = \frac{E}{R_L + R_i}$$

$$\text{load voltage, } V_L = I_L R_L = \frac{E \times R_L}{R_L + R_i} = E - I_L R_i$$

$$\text{percentage voltage regulation} = \frac{V_{NL} - V_{FL}}{V_{FL}} \times 100\%$$

$$\text{load power, } P_L = I_L{}^2 R_L = \frac{V^2{}_L}{R_L} = I_L V_L = \frac{E^2 R_L}{(R_L + R_i)^2}$$

for maximum power transfer, $R_L = R_i$ (matching).

$$\text{maximum load power, } P_{L\max} = \frac{E^2}{4 \ R_L} = \frac{E^2}{4 \ R_i}$$

$$\text{percentage efficiency} = \frac{P_L}{P_T} \times 100\%.$$

☐ Sources in Series-Aiding

$$E_T = E_1 + E_2 + E_3 + \cdots + E_N$$
$$R_{iT} = R_{i1} + R_{i2} + R_{i3} + \cdots + R_{iN}$$

For identical sources,

$$E_T = NE \text{ and } R_{iT} = NR_i.$$

☐ Constant Current Source

$$\text{constant current, I} = \text{short-circuit terminal current}$$
$$= E/R_i$$

$$\text{load current, } I_L = I \times \frac{R_i}{R_L + R_i}$$

☐ Sources in ParallelAiding

$$I_T = I_1 + I_2 + I_3 + \cdots + I_N$$

$$R_{iT} = \cfrac{1}{\cfrac{1}{R_{i1}} + \cfrac{1}{R_{i2}} + \cfrac{1}{R_{i3}} + \cdots + \cfrac{1}{R_{iN}}}$$

For identical sources,

$$I_T = NI \qquad R_{iT} = \frac{R_i}{N}$$

$$E_T = E \text{ (open-circuit voltage of one cell)}$$

5
Magnetic Circuits

IN OUR DISCUSSION of magnetism, electromagnetism, and electromagnetic induction, you will learn

☐ About Faraday's Law, which concerns the creation of electrical energy from mechanical energy.

☐ About the relationship between magnetic flux and magnetic flux density.

☐ About the relationships between magnetomotive force, magnetic field intensity, and flux density when magnetizing a specimen of iron.

☐ To use Rowland's Law, in which the reluctance of an iron specimen is equal to the ratio of the magnetomotive force to the flux.

☐ How the reluctance is related to the permeability of the iron.

☐ About the effect of hysteresis on the magnetizing and demagnetizing of an iron specimen.

The total number of lines in a magnetic field is called the *magnetic flux*. The lines form a complete closed path that is referred to as a magnetic circuit. Figure 5-1 shows a simple magnetic circuit in which the iron ring provides the path for the magnetic flux. The mean length of the magnetic path is ℓ meters and the cross-sectional area of the ring is A square meters.

MAGNETIC FLUX

The S.I. unit of magnetic flux is the *weber* (Wilhelm Edward Weber, 1804–91) and its letter symbol is the Greek letter, ϕ (phi). The weber (Wb) is defined in terms of *Faraday's Law of Electromagnetic Induction;* this law can be stated as follows:

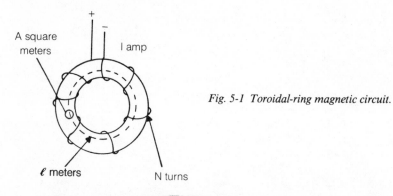

Fig. 5-1 Toroidal-ring magnetic circuit.

When a conductor cuts or is cut by a magnetic flux, an EMF is induced in the conductor. The direction of the induced EMF depends on the direction of the magnetic field and on the direction in which the field moves relative to the conductor. The magnitude of the EMF is proportional to the rate at which the conductor cuts or is cut by the magnetic flux. Therefore, the induced EMF, is given by

$$E \text{ (volts)} = \frac{\text{flux, } \phi \text{ (webers)}}{\text{time, t (seconds)}}$$

The weber is the magnetic flux which, when cut by a conductor in one second, generates an induced EMF of one volt. In the electromagnetic cgs system, the flux unit is the *line* or *maxwell* in which 1 Weber = 10^8 maxwells.

FLUX DENSITY

In Fig. 5-1, the magnetic flux has a cross-sectional area of A square meters; the flux density, B, is therefore defined by

$$\text{flux density, } B = \frac{\text{flux, } \phi \text{ (webers)}}{\text{cross-sectional area, A (square meters)}}.$$

The SI unit of flux density is the tesla (T) so that 1 T = 1 Wb/m². In the cgs system, the flux density unit is the gauss (maxwell/cm²), which is equivalent to 1×10^{-4} tesla.

Faraday's Law relates to the generator effect in which mechanical energy is converted into electrical energy. The reverse action is the *motor effect* (Fig. 5-2) in which a force is exerted on a current-carrying conductor, situated in a magnetic field. In SI units: force (newtons) = flux density, B (tesla) × current, I (amperes) × conductor's length, ℓ (meters).

Referring again to Faraday's Law (Fig. 5-3) the induced EMF is E (volts) = flux density, B (teslas) × conductor's length, ℓ (meters) × conductor's velocity, v (meters per second)

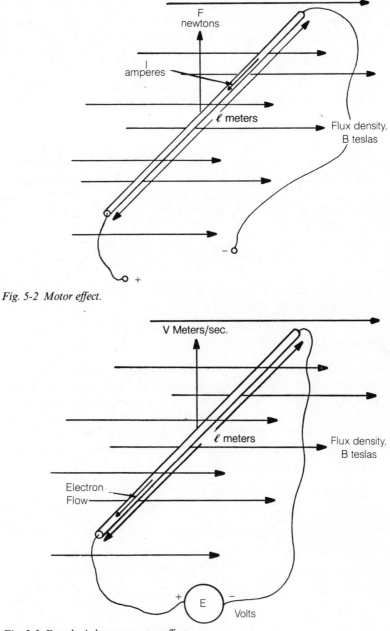

Fig. 5-2 Motor effect.

Fig. 5-3 Faraday's law: generator effect.

MAGNETO-MOTIVE FORCE

The magnetic flux in the iron ring of Fig. 5-1 is established as the result of the current, I, flowing through the exciting coil consisting of N turns. Increasing the current and the number of turns increases the flux. This is equivalent to increasing the EMF applied across a resistor and observing that the current increases. The product of the current and the number of turns is therefore called the *magnetomotive force* (MMF), $\mathscr{F}$. The SI unit of the MMF is the ampere-turn or ampere because the number of turns has no dimensions. Therefore magnetomotive force, $\mathscr{F}$ (amperes) = current, I (amperes) $\times$ number of turns, N. The cgs unit of MMF is the gilbert, in which 1 ampere-turn = $4\pi/10$ gilberts.

MAGNETIC FIELD INTENSITY

In the case of the iron ring example, the MMF is distributed over a magnetic path whose mean length is ℓ meters. The MMF per meter length of the magnetic circuit is called the *magnetic-field intensity* or *magnetizing force*, H. The SI unit of magnetic field intensity is the ampere-turn per meter or ampere per meter so that

$$\text{magnetic field intensity, H} = \frac{\text{current, I (amperes)} \times \text{number of turns, N}}{\text{length, } \ell \text{ (meters).}}$$

In the cgs system, the unit of magnetic field intensity is the oersted, in which 1 ampere-turn per meter = $4\pi \times 10^{-3}$ oersted.

PERMEABILITY OF FREE SPACE

In Fig. 5-4, X represents the cross-section of a long straight conductor that is positioned in a vacuum and carries an electron current of one ampere emerging from the paper. The magnetic flux surrounding X will be in the shape of concentric circles. One such circle, marked C, has a radius of 1 meter and represents the path of one line of flux. The MMF acting on this path is one ampere and because the path length is 2π meters, the magnetic field intensity is $1/2\pi$ ampere per meter.

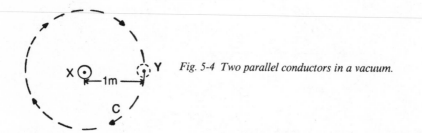

Fig. 5-4 *Two parallel conductors in a vacuum.*

If B is the flux density in teslas around the circle, C, and Y is a second conductor that is parallel to X and also carries a current of one ampere, then force exerted on Y = B (teslas) $\times$ 1 (ampere) $\times$ 1 (meter).

However, from the definition of the ampere in Chapter 2, this force must be 2×10^{-7} newton. Therefore B = 2×10^{-7} tesla and

$$\frac{\text{flux density at Y, B}}{\text{magnetic field intensity at Y, H}} = \frac{2 \times 10^{-7}\ \text{T}}{1/2\ \pi\ \text{A/m}}$$

$$= 4\ \pi \times 10^{-7}$$
$$= 12.57 \times 10^{-7}\ \text{SI units}$$

The ratio of B/H for a vacuum is called the *permeability of free space* whose letter symbol is μ_0; consequently $B = \mu_0\ H$. The value of μ_0 is determined by the system of units; for example, in the cgs system the value of μ_0 is 1.

Many substances are nonmagnetic and have the same permeability as free space. However, there are a few paramagnetic materials such as aluminum and platinum that have a permeability slightly higher than that of free space. Yet other elements, for example, copper and silver, are diamagnetic and have a permeability slightly less than free space. Finally, the ferromagnetic substances (iron, nickel, cobalt) have permeabilities many times greater than free space.

Example 5-1

A coil of 500 turns carries a current of 6 A and is wound uniformly over a nonmagnetic toroidal ring with a mean circumference of 40 cm and a cross-sectional area of 3 cm². Calculate the magnetic field intensity, the flux density, and the total flux.

Solution

$$\text{magnetomotive force, } \mathscr{F} = 6\ \text{A} \times 500 = 3000\ \text{ampere-turns}$$
$$\text{length of magnetic path, } \ell = 40\ \text{cm} = 0.4\ \text{m}$$

$$\text{magnetic field intensity, } H = \frac{\mathscr{F}}{\ell} = \frac{3000}{0.4} = 7500\ \text{amperes/meter}$$

$$\text{flux density, } B = \mu_0\ H$$
$$= 12.57 \times 10^{-7} \times 7500$$
$$= 0.00943\ \text{tesla}$$
$$\text{cross-sectional area, } A = 3\ \text{cm}^2$$
$$= 3 \times 10^{-4}\text{m}^2$$

$$\text{total flux, } \phi = B \times A$$
$$= 0.00943\text{T} \times 3 \times 10^{-4}\text{m}^2$$
$$= 2.83\ \mu\text{Wb}$$

Example 5-2

An airgap is 5 mm long and has a cross-sectional area of 30 cm². What is the number of ampere-turns required to establish a total flux of 8.0 mWb in the airgap?

Solution

$$\text{cross-sectional area, } A = 30 \times 10^{-4}\text{m}^2$$

$$\text{flux density, B} = \frac{\phi}{A} = \frac{8.0 \times 10^{-3}}{30 \times 10^{-4}} = 2.67 \text{ T}$$

$$\text{magnetic field intensity, H} = \frac{B}{\mu_0} = \frac{2.67}{12.57 \times 10^{-7}}$$

$$= 2.124 \times 10^6 \text{ A.T.}$$

$$\text{length of magnetic path, } \ell = 5 \times 10^{-3}\text{m}$$
$$\text{total ampere-turns} = 2.124 \times 10^6 \times 5 \times 10^{-3}$$
$$= 10,620$$

RELATIVE PERMEABILITY

It is well known that the magnetic flux inside a coil is intensified when a soft iron core is inserted. Consequently, if the magnetic core of the toroidal ring in Fig. 5-1 is replaced by an iron core, the flux for a given MMF is greatly increased. The ratio of the flux density established in the core material to the flux density produced in a vacuum (or nonmagnetic core) for the same magnetic field intensity, is called the *relative permeability, μ_r*. For air $\mu_r = 1$, but for certain nickel-iron alloys, the relative permeability can be as high as 100,000. It should be emphasized that the value of μ_r depends on the magnitude of the magnetic field intensity and is not a constant for a particular alloy (Fig. 5-5).

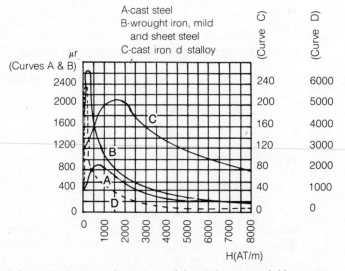

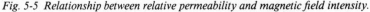

Fig. 5-5 Relationship between relative permeability and magnetic field intensity.

For a core material with a relative permeability, μ_r

$$B = \mu_0 \mu_r H$$

$$\text{absolute permeability, } \mu = \frac{B}{H} = \mu_0 \mu_r$$

Referring back to Fig. 5-1, the iron ring has a cross-sectional area of A square meters and a mean circumference of ℓ meters. It is wound with N turns of a coil in which a current of I amperes flows. Then: total flux, ϕ = flux density, B × area, A. Magnetomotive force, $\mathscr{F}$ = magnetic field intensity, H × length, ℓ.

$$\frac{\text{MMF, } \mathscr{F}}{\text{flux, } \phi} = \frac{H\ell}{BA} = \frac{H\ell}{\mu_0 \, \mu_r \, HA} = \frac{\ell}{\mu_0 \, \mu_r \, A}$$

This relationship is sometimes known as Rowland's Law and is the magnetic equivalent of the Ohm's Law electrical equation.

$$\frac{\text{EMF, } E}{\text{current, } I} = \text{resistance, } R = \frac{\text{resistivity, } \rho \times \text{length, } \ell}{\text{cross-sectional area, } A}$$

The quantity

$$\frac{\ell}{\mu_0 \, \mu_r \, A}$$

is therefore comparable to resistance and is called the *reluctance*, $\mathscr{R}$, of the magnetic circuit. There are no special SI units for reluctance that can be referred to either as ampere-turns per weber or as the reciprocal of the henry, which is the unit of inductance (Chapter 6). The reciprocal of reluctance is the *permeance*, $\mathscr{P}$, whose SI unit is the henry or weber per ampere-turn.

In the cgs system, the unit of reluctance is the *rel*, which is equivalent to one gilbert per maxwell; 1 SI unit of reluctance = $4 \, \pi \times 10^{-9}$ rels.

In comparing units

$$R = \frac{\rho\ell}{A} = \frac{\ell}{\sigma A}$$

and

$$\mathscr{R} = \frac{\ell}{\mu_0 \, \mu_r \, A} = \frac{\ell}{\mu A}$$

the absolute permeability μ, corresponds to the conductivity, σ; the SI unit of permeability is the henry per meter.

Example 5-3

A particular magnetic circuit has an air gap that is 7 mm long and has a cross-sectional area of 30 cm². Calculate the reluctance of the gap and the MMF required to establish a flux of 900 μWb across the gap.

Solution

$$\text{reluctance of the gap, } \mathscr{R} = \frac{\ell}{\mu_0 \, A} = \frac{7 \times 10^{-3} \text{ m}}{12.57 \times 10^{-7} \times 30 \times 10^{-4} \text{ m}^2}$$
$$= 1.86 \times 10^6 \text{ ampere-turns per weber}$$

$$\text{required MMF} = \mathcal{R} \times \phi = 1.86 \times 10^6 \times 900 \times 10^{-6}$$
$$= 1670 \text{ ampere-turns}$$

Example 5-4

An iron ring has a mean circumference of 60 cm and a uniform cross-sectional area of 5 cm². Corresponding to a flux density of 1.4 Wb/m², the relative permeability of the iron is 2400. For the value of this flux density, calculate the reluctance of the circuit and the MMF is required.

Solution

$$\text{reluctance of the iron ring, } \mathcal{R} = \frac{\ell}{\mu_0 \, \mu_r \, A}$$

$$= \frac{60 \times 10^{-2}}{12.57 \times 10^{-7} \times 2400 \times 5 \times 10^{-4}}$$

$$= 3.98 \times 10^5 \text{ ampere-turns per weber}$$

$$\text{flux, } \phi = BA = 1.4 \times 5 \times 10^{-4} = 7 \times 10^{-4} \text{ Wb.}$$

$$\text{required MMF} = \mathcal{R} \times \phi = 3.98 \times 10^5 \times 7 \times 10^{-4} = 280 \text{ ampere-turns}$$

SERIES MAGNETIC CIRCUIT

Figure 5-6 shows a composite magnetic circuit consisting of three specimens of iron and an air gap. It is assumed that the same flux is established throughout the circuit and that the magnetic leakage is negligible. You can then apply the principles used in analyzing a series electrical circuit to the composite magnetic circuit. Therefore: total reluctance, $\mathcal{R}_T = \mathcal{R}_1 + \mathcal{R}_2 + \mathcal{R}_3 + \mathcal{R}_4$. Total MMF, $\mathcal{F}_T = \mathcal{F}_1 + \mathcal{F}_2 + \mathcal{F}_3 + \mathcal{F}_4$. There are comparable similarities between the equations for parallel electrical and magnetic circuits (see Example 5-7).

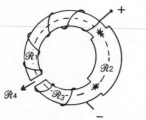

Fig. 5-6 Composite series magnetic circuit.

One important difference between electrical and magnetic circuits lies in the concept of power. Although energy must be supplied continuously to an electrical circuit to sustain the flow of electricity, the magnetic flux, once set up, does not need any further energy. In the case of the magnetized iron ring, the only energy supplied after the flux is fully established is dissipated by the coil's resistance in the form of heat.

MAGNETIZATION CURVE

Because the reluctance for a particular type of iron depends on the flux density and is not a constant, it is common practice to use the magnetization curve to obtain directly the corresponding values of the flux density, B, and the magnetic field intensity, H. This curve is derived experimentally and illustrates the relationship between B and H for a given specimen of iron; Figure 5-7A shows some examples. For each curve there is an initial sharp rise followed by a levelling off caused by partial saturation of the iron.

In Fig. 5-7B, OAC represents the magnetization curve for a certain type of iron. If after reaching a maximum value corresponding to OK, the magnetic field intensity is reduced (by decreasing the current in the exciting coil), the flux density follows the curve CD. When the field intensity, H, is reduced

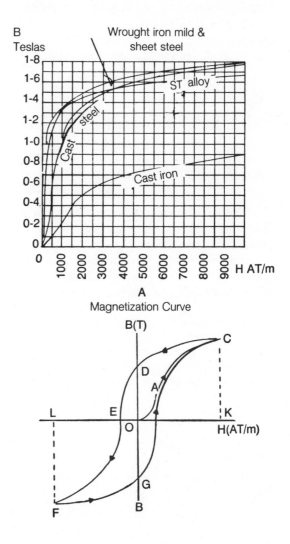

Fig. 5-7 Comparison of the magnetization curve and the hysteresis loop.

to zero, the value of B still present in the iron is represented by OD, which is called the *residual* or *remanent flux density*. If the current is now reversed so that the field intensity is increased in the opposite direction, there will be a value OE for which the magnetic flux density is reduced to zero; this value of H is called the *coercive force*. Further increasing the field intensity raises the flux density in the reverse direction (as represented by the curve EF).

HYSTERESIS LOOP

If the field intensity is now changed from OL back to OK, the flux density follows the curve FGC, which is similar to the curve CDEF. The complete closed figure is called the *hysteresis loop;* the word *hysteresis* indicates that the result (the flux density, B) lags behind the cause (the field intensity, H). Hysteresis produces a heating effect in the iron and therefore represents an energy loss from the source of the exciting coil. The amount of energy loss in joules per cubic meter is represented by the area of the loop. This is indicated in the Steinmetz equation:

$$\text{hysteresis power loss} = kfvB_{max}^{1.6} \text{ watts}$$

where

k = a constant for a particular specimen of iron
f = frequency in hertz
v = volume of iron in cubic meters
B_{max} = maximum flux density in teslas corresponding to the peak field intensity in ampere turns per meter

For most ferromagnetic materials, the ratio of the residual flux density to the maximum flux density is about 0.7. However, the coercive force can range from more than 10,000 to less than 10 ampere-turns per meter.

The easiest way to demagnetize the iron specimen and return to the point 0 is to reverse the exciting current many times while gradually reducing the value of the peak current to zero.

Example 5-5

A cast-iron ring has a mean circumference of 50 cm and a cross-sectional area of 4 cm^2. What is the current required in a coil of 300 turns to establish a total flux of 160 μWb?

Solution

$$\text{flux density, B} = \frac{\phi}{A} = \frac{160 \times 10^{-6}}{4 \times 10^{-4}} = 0.4 \text{ Wb/m}^2$$

Using Fig. 5-7A, the magnetic field intensity, H, corresponding to a flux density of 0.4 Wb/m^2 in cast-iron, is 1500 ampere-turns per meter.

$$\text{required MMF, F} = H \times \ell = 1500 \times 50 \times 10^{-2}$$
$$= 750 \text{ ampere-turns}$$

$$\text{magnetizing current} = \frac{750}{300} = 2.5 \text{ A}$$

Example 5-6

Figure 5-8 represents a cast steel ring with a cross-sectional area of 6 cm^2 and a mean circumference of 55 cm. There are two radial cuts at diametrically opposite points, and each of these are filled with nonmagnetic material to a thickness of 0.8 mm. Assuming no magnetic leakage, calculate the total MMF required to establish a flux density of 1.5 T.

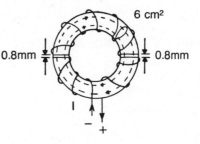

Fig. 5-8 Diagram for example 5-6.

Solution

From Fig. 5-7A, the number of ampere-turns per meter required for the cast steel is 3000. The corresponding number of ampere-turns is 3000 × 55 × 10^{-2} = 1650.

For the nonmagnetic gaps the magnetic field intensity is

$$H = \frac{1.5}{12.57 \times 10^{-7}}$$

$$= 1.19 \times 10^6 \text{ ampere-turns per meter}$$

total ampere-turns for the two gaps = 2 × 1.19 × 10^6 × 0.8 × 10^{-3} = 1900

total MMF required = 1650 + 1900 = 3550 ampere-turns

Example 5-7

Figure 5-9 shows a cast-iron magnetic core such as those used in transformers. The cross-sectional area of each side limb is 10 cm^2 while the center limb's area is 13 cm^2. Neglecting any magnetic leakage, calculate the MMF required to establish a flux density of 0.2 T in the airgap.

Fig. 5-9 Diagram for example 5-7.

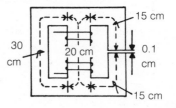

Solution

This magnetic arrangement is equivalent to a series-parallel circuit. The right-side limb is in series with the airgap, and their total reluctance is in parallel with the left-side limb. Finally, this combination is in series with the center limb.

From Fig. 5-7A, a flux density of 0.2 T in cast iron is produced by a magnetic field intensity of 800 ampere-turns per meter. Therefore, the ampere turns for the iron in the right-hand limb is $800 \times 0.3 = 240$. The magnetic field intensity for the airgap is

$$\frac{B}{\mu_0} = \frac{0.2}{12.57 \times 10^{-7}} = 1.6 \times 10^{-5}$$

and the corresponding ampere turns are

$$1.6 \times 10^{-5} \times 0.1 \times 10^{-2} = 160$$

The total ampere turns for the left-hand limb is

$$240 + 160 = 400$$

which will result in a magnetic field intensity of

$$\frac{400}{30 \times 10^{-2}} = 1333 \text{ ampere turns per meter}$$

Again, using Fig. 5-7A, the corresponding flux density is 0.375 T. Therefore, the total flux in the center limb is

$$(0.2 + 0.375) \times 10 \times 10^{-4} = 5.75 \times 10^{-4} \text{ Wb}$$

and the flux density is

$$\frac{5.75 \times 10^{-4}}{13 \times 10^{-4}} = 0.44 \text{ T}$$

which is produced by a magnetic field intensity of 1750 ampere-turns per meter. The ampere-turns for the center limb are $1750 \times 20 \times 10^{-2} = 350$, and the total MMF required is $400 + 350 = 750$ ampere turns.

CHAPTER SUMMARY

Table 5-1 shows a comparison between electric and magnetic units.

□ Equations

$$\text{MMF, } \mathscr{F} = IN \text{ ampere-turns}$$

$$\text{magnetic field intensity, } H = \frac{IN}{\ell} \text{ ampere-turns/m}$$

$$\text{flux density, } B = \frac{\phi}{A} \text{ tesla or Wb/m}^2$$

$$\text{permeability of free space} = 4\pi \times 10^{-7} = 12.57 \times 10^{-7} \text{ SI units}$$

Table 5-1. Comparison between Electric and Magnetic Units

Electric Circuit Ohm's Law, Resistance = EMF/Current		Magnetic Circuit Rowland's Law, Reluctance = MMF/Flux	
Quantity	SI Unit	Quantity	SI Unit
EMF	Volt	MMF	ampere-turn
Current	ampere	Flux	weber
Current density	ampere per square meter	flux density	weber per square meter or telsa
Resistance	Ohm	Reluctance	ampere-turn per weber
Conductance	siemens	Permeance	henry or weber per ampere-turn
Conductitvity	siemens per meter	Permeability	henry per meter
Electric field intensity	volts per meter	Magnetic field intensity	ampere-turns or amperes per meter

absolute permeability, $\mu = \mu_0 \mu_r$

for nonmagnetic materials, $B = \mu_0 H$

for magnetic materials, $B = \mu_0 \mu_r H$

reluctance, $\mathscr{R} = \dfrac{\ell}{\mu A} = \dfrac{\ell}{\mu_0 \mu_r A}$ ampere-turn perweber

permeance, $\mathscr{P} = \dfrac{1}{\mathscr{R}} = \dfrac{\mu A}{\ell}$ henry

☐ Series Magnetic Circuit

$$\mathscr{R}_T = \mathscr{R}_1 + \mathscr{R}_2 + \mathscr{R}_3 + \cdots + \mathscr{R}_N$$

☐ Parallel Magnetic Circuit

$$\frac{1}{\mathscr{R}_T} = \frac{1}{\mathscr{R}_1} + \frac{1}{\mathscr{R}_2} + \frac{1}{\mathscr{R}_3} + \cdots + \frac{1}{\mathscr{R}_N}$$

6

Inductance

IN STUDYING THE effects of self-inductance, you will learn:

☐ How to compare the electrical properties of resistance and inductance.

☐ The definition of the unit of inductance.

☐ Which physical factors determine a coil's self-inductance.

☐ How to calculate the amount of energy stored in the magnetic field surrounding an inductor.

☐ What occurs when a number of inductors are connected in series.

☐ What occurs when a number of inductors are connected in parallel or series-parallel.

☐ About the concept of the time constant, L/R seconds, and its relation to the duration of the transient state.

Inductance is that property of an electrical circuit which opposes any *sudden change* of current. This compares with the property of resistance which only limits or opposes the flow of current. The circuit of Fig. 6-1 illustrates the differences between the two properties. Although a straight conductor possesses inductance, the property is most marked in a coil, which is referred to as an *inductor* with L as the letter symbol.

INTRODUCTION TO INDUCTANCE

As soon as the switch, S, is closed, the current in the resistor immediately jumps from zero to a value of E/R amperes. The potential difference or voltage drop across R exactly balances (opposes) the applied voltage, E. It is therefore appropriate to write $V_R = -I_R \times R$. By Kirchhoff's Voltage Law, $E + V_R = 0$ which leads to $E = I_R \times R$.

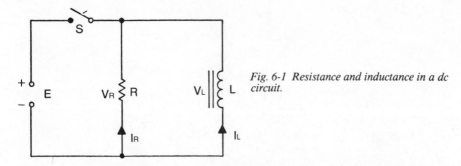

Fig. 6-1 Resistance and inductance in a dc circuit.

By contrast, as soon as the current starts to grow in the coil, a magnetic flux is created, and then as this field expands out, it cuts the turns of the coil. By Faraday's Law, a counter EMF, V_L, is self-induced into the coil, and this voltage must balance exactly (oppose) the applied EMF.

Because the counter EMF is produced by a moving flux, this requires a changing current (a constant flux would generate zero induced voltage). The current therefore starts at zero and has a rate of growth that is measured in amperes per second. Summarizing the foregoing, the current in the coil is zero before the switch is closed and remains at zero just after the switch is closed; in other words, inductance opposes any sudden change in the current's value.

The size of the counter EMF is determined by the rate of cutting of the magnetic flux. This in turn depends on the rate of change of current and a factor related to the coil. This factor is the inductance that is associated with the coil's number of turns, nature of the core, cross-sectional area, and length. In equation form,

$$V_L = -L \times \frac{\Delta I_L}{\Delta T}$$

where

V_L = the counter EMF, with the negative sign indicating that V_L opposes E

$\dfrac{\Delta I_L}{\Delta T}$ = the rate of change of current in amperes per second

The symbol "Δ" means "the change in" so that ΔI_L is the change in current and ΔT the corresponding change in time. L = the coil's inductance measured in henrys (H).

The inductance equals one henry (after Joseph Henry, 1797–1878) if, when the current is changing at the rate of one ampere per second, the induced EMF is one volt. By Kirchhoff's Voltage Law,

$$E + V_L = 0$$

Then

$$E = L \times \frac{\Delta I_L}{\Delta T}$$

or rate of change of current,

$$\frac{\Delta I_L}{\Delta T} = \frac{E}{L}$$

amperes per second.

The current in the inductor therefore starts at zero and grows at a constant rate. However, this neglects the coil's resistance, R_L. In practice, I_L takes an appreciable time to reach a final steady value equal to E/R_L amperes. When this final condition has been reached, the flux is stationary and the counter EMF is zero. If S is now opened, there is no immediate change in the inductor's current. However, as I_L subsequently begins to decay, the magnetic flux starts to collapse and induces a counter EMF whose polarity is such as to oppose the decrease in I_L. This is an example of *Lenz's Law*, which states that the polarity of the induced EMF is such as to oppose the change producing the EMF. The return path for I_L is through the resistor so that the current through both components is now the same and will again take an appreciable time to decrease to zero.

Example 6-1

The current through an inductor changes from 7 A to 2 A in a time of 25 milliseconds. If the self-inductance is 1.5 henry, what is the average value of the induced EMF?

Solution

change in current, $\Delta I = 7\text{ A} - 2\text{ A} = 5\text{ A}$
corresponding change in time, $\Delta T = 25 \times 10^{-3}$ seconds

$$\text{average induced EMF} = L \times \frac{\Delta I}{\Delta T} = 1.5\text{ H} \times \frac{5\text{ A}}{25 \times 10^{-3}\text{ s}}$$

$$= 300 \text{ volts}$$

FACTORS DETERMINING A COIL'S INDUCTANCE

Consider a coil of N turns with a length of ℓ meters and a cross-sectional area of A square meters (Fig. 6-2). The current in this coil changes by ΔI amperes in a corresponding time change of ΔT seconds. This coil is uniformly wound on a magnetic core with a relative permeability, μ_r. Using Faraday's Law and regarding each turn as a conductor,

$$E = N\frac{\Delta\phi}{\Delta T} = L\frac{\Delta I}{\Delta T}$$

where $\Delta\phi$ in webers is the change in flux associated with the change of current, ΔI. Therefore,

$$L = N\frac{\Delta\phi}{\Delta I} = \frac{\text{change in flux-linkages}}{\text{change of current}}$$

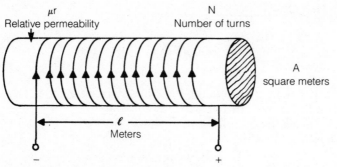

Fig. 6-2 Factors effecting the self-inductance of a coil.

The term *flux linkages* means the product of the flux and the number of turns through which the flux passes or with which the flux is linked. This leads to an alternative definition of the henry. The inductance of a coil is one henry if a current change of one ampere produces a change in flux-linkages of 1 weber-turn. Therefore,

$$L = N \frac{\Delta\phi}{\Delta I}$$

$$= NA \times \frac{\Delta B}{\Delta I}$$

$$= NA\mu_o\mu_r \times \frac{\Delta H}{\Delta I}$$

$$= NA\mu_o\mu_r \frac{N}{\ell} \times \frac{\Delta I}{\Delta T}$$

$$= \frac{\mu_o\mu_r N^2 A}{\ell} = \frac{N^2}{\mathscr{R}} \text{ henrys}$$

where $\mathscr{R}$ is the reluctance of the magnetic path

Because the inductance depends on the relative permeability, it is not a constant but varies with the current carried by the coil. The variation of inductance with current is similar in shape to the μ_r/H curves of Fig. 5-5. For a coil with a nonmagnetic core, the inductance is

$$L = \frac{\mu_o N^2 A}{\ell} = 12.57 \times 10^{-7} \times \frac{N^2 A}{\ell} \text{ henrys}$$

Example 6-2

A coil of 500 turns is wound on a nonmagnetic core. A current of 6 A flows through the coil and establishes a flux of 300 μWb. What is the inductance of the coil and the average value of the induced EMF if the current is reversed in 25 milliseconds?

Solution

$$\text{inductance, } L = N \times \frac{\Delta\phi}{\Delta I}$$

$$= 500 \times \frac{300 \times 10^{-6}}{6} \text{ H}$$

$$= 25 \text{ mH}$$

$$\text{average induced EMF, } E = N \times \frac{\Delta\phi}{\Delta T}$$

$$= 500 \times \frac{2 \times 300 \times 10^{-6}}{25 \times 10^{-3}} = 12 \text{ V}$$

$$\text{alternatively, } E = L \times \frac{\Delta I}{\Delta T} = 25 \times 10^{-3} \times \frac{2 \times 6}{25 \times 10^{-3}}$$

$$= 12 \text{ V}$$

Example 6-3

A cast-iron ring has a mean circumference of 50 cm and a cross-sectional area of 4 cm². The ring is wound uniformly with a coil of 300 turns. Calculate the inductance if currents of (a) 2 A and (b) 8 A are reversed in the coil.

Solution

$$\text{(a) magnetic field intensity, } H = \frac{IN}{\ell} = \frac{2 \times 300}{50 \times 10^{-2}}$$

$$= 1200 \text{ ampere-turns per meter}$$

from *Fig. 5-7A,* flux density, B = 0.3T

$$\text{flux, } \phi = BA = 0.3 \times 4 \times 10^{-4} \text{ Wb} = 120 \ \mu\text{Wb}$$

$$\text{inductance, } L = N \times \frac{\Delta\phi}{\Delta I}$$

$$= 300 \times \frac{2 \times 120 \times 10^{-6}}{2 \times 2} \text{ H}$$

$$= 0.018 \text{ H}$$

$$\text{(b) magnetic field intensity, } H = \frac{8 \times 300}{50 \times 10^{-2}}$$

$$= 4800 \text{ ampere-turns per meter}$$

corresponding flux density, B = 0.71 T

$$\text{flux, } \phi = BA = 0.71 \times 4 \times 10^{-4} \text{ Wb} = 284 \ \mu \text{ Wb}$$

$$\text{inductance, } L = 300 \times \frac{2 \times 284 \times 10^{-6}}{2 \times 8} \text{ H}$$

$$= 0.011 \text{ H}$$

This example illustrates the fact that the value of the inductance depends on the current.

Example 6-4

If the cast-iron ring of example 6-3 is replaced by a nonmagnetic core, calculate the new inductance of the coil.

Solution

$$\text{new inductance, } L = \frac{\mu_o N^2 A}{\ell}$$

$$= \frac{12.57 \times 10^{-7} \times (300)^2 \times 4 \times 10^{-4}}{50 \times 10^{-2}} \text{ H}$$

$$= 90.5 \ \mu H$$

This shows that a coil with an inductance of the order of henrys will require some form of iron core.

ENERGY STORED IN THE MAGNETIC FIELD SURROUNDING AN INDUCTOR

The magnetic field associated with an inductor represents a form of energy that is created by the current flowing through the coil. The energy is delivered by the source from which the current is drawn. Assume that the coil has a constant inductance of L henrys and that after a switch is closed, the current grows at a constant rate from zero to I amperes in a time of T seconds (Fig. 6-3). The average current is I/2 and the source voltage, E, is LI/T.

Average power delivered to the magnetic field equals

$$E \times \frac{I}{2} = \frac{LI}{T} \times \frac{I}{2} = \frac{LI^2}{2T} \text{ watts}$$

total energy stored in magnetic field = average power × time

$$= \frac{LI^2}{2T} \times T$$

$$= \frac{1}{2} LI^2 \text{ joules}$$

If the switch is now opened, the current decays and the energy in the magnetic field dissipates in the form of the arc that occurs between the contacts of the switch.

Example 6-5

A coil with an inductance of 4 H and a resistance of 25 Ω is connected across a 60-V source. Calculate (a) the initial rate of current growth, (b) the value of the final current and (c) the energy stored in the magnetic field when the final conditions have been reached.

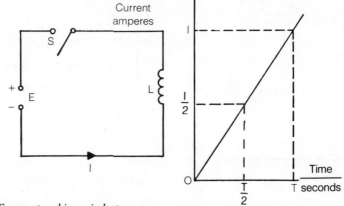

Fig. 6-3 Energy stored in an inductor.

Solution

(a) initial rate of growth of current $= \dfrac{E}{L}$

$$= \dfrac{60 \text{ V}}{4 \text{ H}}$$

$$= 15 \text{ amperes per second}$$

(b) final current $= \dfrac{E}{R} = \dfrac{60 \text{ V}}{25 \text{ } \Omega} = 2.4 \text{ A}$

You might have some difficulty in reconciling the answers to (a) and (b) until you realize that the current reaches its final value in less than one second.

(c) energy stored $= \frac{1}{2}LI^2 = \frac{1}{2} \times 4 \times (2.4)^2 = 11.5 \text{ joules}$

INDUCTORS IN SERIES AND PARALLEL

Referring to Fig. 6-4, the inductors L_1, L_2, and L_3 are in series so that the same rate of change of current is associated with each inductor. Therefore

$$V_1 = -L_1 \frac{\Delta I}{\Delta T}, \; V_2 = -L_2 \frac{\Delta I}{\Delta T}, \; V_3 = -L_3 \frac{\Delta I}{\Delta T}$$

and

$$E + V_1 + V_2 + V_3 = 0$$

This leads to

$$E = \frac{\Delta I}{\Delta T} (L_1 + L_2 + L_3)$$

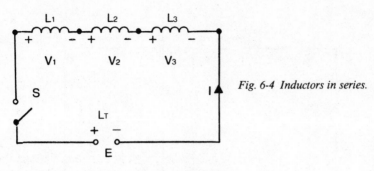

Fig. 6-4 Inductors in series.

If L_T is the total equivalent inductance,

$$E = L_T \frac{\Delta I}{\Delta T}$$

so that

$$L_T = L_1 + L_2 + L_3$$

This is similar to the formula for resistors in series.

In Fig. 6-5, the inductors L_1, L_2, and L_3 are in parallel so that the total rate of change of current associated with the source is equal to the sum of the rates of change of current in the three branches. Then

$$\frac{\Delta I_1}{\Delta T} = \frac{E}{L_1}, \frac{\Delta I_2}{\Delta T} = \frac{E}{L_2}, \frac{\Delta I_3}{\Delta T} = \frac{E}{L_3}$$

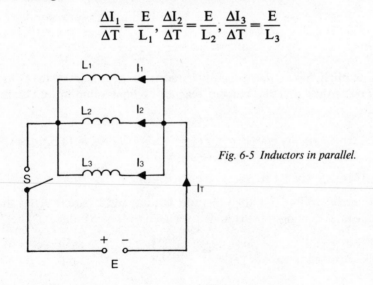

Fig. 6-5 Inductors in parallel.

and

$$\frac{\Delta I_T}{\Delta T} = \frac{\Delta I_1}{\Delta T} + \frac{\Delta I_2}{\Delta T} + \frac{\Delta I_3}{\Delta T} = E \times \left(\frac{1}{L_1} + \frac{1}{L_2} + \frac{1}{L_3} \right)$$

If L_T is the total equivalent inductance

$$\frac{\Delta I_T}{\Delta T} = \frac{E}{L_T} = E \times \left(\frac{1}{L_1} + \frac{1}{L_2} + \frac{1}{L_3} \right)$$

and

$$\frac{1}{L_T} = \frac{1}{L_1} + \frac{1}{L_2} + \frac{1}{L_3}$$

The same type of reciprocal formula applies to resistors in parallel.

Example 6-6

In Fig. 6-4, $L_1 = 1.8$ H, $L_2 = 3.7$ H, $L_3 = 2.4$ H, and $E = 38$ V. Calculate the voltages V_1, V_2, V_3, and the rate of growth of current. Determine the energy stored in each inductor after 5 seconds from the time S is closed.

Solution

total inductance, $L_T = L_1 + L_2 + L_3 = 1.8 + 3.7 + 2.4 = 7.9$ H

$$\text{rate of growth of current} = \frac{38 \text{ V}}{7.9 \text{ H}} = 4.81 \text{ amperes per second}$$

$$V_1 = 4.81 \times 1.8 = 8.66 \text{ V}$$
$$V_2 = 4.81 \times 3.7 = 17.80 \text{ V}$$
$$V_3 = 4.81 \times 2.4 = 11.54 \text{ V}$$

Check: $V_1 + V_2 + V_3 = 8.66 + 17.80 + 11.54 = 38$ V. After 5 seconds the circuit current, I, is $4.81 \times 5 = 24.05$ A.

Energy stored in the 1.8 H inductor $= \frac{1}{2} \times 1.8 \times (24.05)^2 = 521$ joules.

Energy stored in the 3.7 H inductor $= \frac{1}{2} \times 3.7 \times (24.05)^2 = 1070$ joules.

Energy stored in the 2.4 H inductor $= \frac{1}{2} \times 2.4 \times (24.05)^2 = 694$ joules.

Total energy stored is $521 + 1070 + 694 = 2285$ joules.

Check: total energy stored $= \frac{1}{2} \times L_T \times I^2 = \frac{1}{2} \times 7.9 \times (24.05)^2 = 2285$ joules.

Example 6-7

In Fig. 6-5, $L_1 = 1.8$ H, $L_2 = 3.7$ H, $L_3 = 2.4$ H, and $E = 38$ V. Calculate the rates of current growth in the individual inductors and the total rate of current growth. Determine the energy stored in each inductor after 5 seconds from the time S is closed.

Solution

Total inductance, L_T, is given by

$$\frac{1}{L_T} = \frac{1}{1.8} + \frac{1}{3.7} + \frac{1}{2.4} = 0.555 + 0.270 + 0.417 = 1.242$$

$$L_T = \frac{1}{1.242} = 0.805 \text{ H}$$

Total rate of current growth is

$$\frac{38}{0.850} = 47.2 \text{ amperes per second}$$

Rate of current growth in the 1.8 H inductor is

$$\frac{38}{1.8} = 21.1 \text{ amperes per second}$$

Rate of current growth in the 3.7 H inductor is

$$\frac{38}{3.7} = 10.3 \text{ amperes per second}$$

Rate of current growth in the 2.4 H inductor is

$$\frac{38}{2.4} = 15.8 \text{ amperes per second}$$

Check: total rate of current growth = 21.1 + 10.3 + 15.8 = 47.2 amperes per second.

After 5 seconds, the currents in the individual inductors are 21.1 × 5 = 105.5 A, 10.3 × 5 = 51.5 A, 15.8 × 5 = 79.0 A, and the total current is 47.2 × 5 = 236.0 A.

Energy stored in the 1.8 H inductor is

$$\frac{1}{2} \times 1.8 \times (105.5)^2 = 10017 \text{ joules}$$

Energy stored in the 3.7 H inductor is

$$\frac{1}{2} \times 3.7 \times (51.5)^2 = 4907 \text{ joules}$$

Energy stored in the 2.4 H inductor is

$$\frac{1}{2} \times 2.4 \times (79.0)^2 = 7489 \text{ joules}$$

Total energy stored is

$$10017 + 4907 + 7489 = 22415 \text{ joules}$$

Check: Total energy stored is ½ × 0.805 × (236)² = 22415 joules.

TIME CONSTANT OF AN INDUCTOR AND A RESISTOR IN SERIES

Immediately after the switch S of Fig. 6-6 is closed in position 1, the inductance prevents any sudden change of current so that the initial circuit conditions are

$$V_L = E, \; V_R = 0, \; I = 0$$

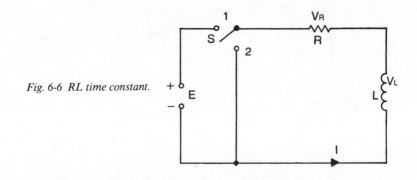

Fig. 6-6 RL time constant.

and the rate of growth of current is E/L amperes per second.

As the current increases, V_R rises, and therefore V_L falls so that the rate of current growth is less.

The time constant of the circuit is L/R seconds, where L is the inductance in henrys and R is the resistance in ohms. This is the time taken for the current to reach 63.2% of its final steady-state value, which is limited only by the resistance, R. You can estimate the order of the time constant's value from the values shown in Table 6-1.

Table 6-1. LR Time Constants.

L	R	Time constant L/R
H	Ω	seconds
H	kΩ	milliseconds
H	MΩ	microseconds
mH	Ω	milliseconds
mH	kΩ	microseconds
mH	MΩ	nanoseconds
μH	Ω	microseconds
μH	kΩ	nanoseconds
μH	MΩ	picoseconds

The equations relating I, V_R, V_L, and time, are

$$I = \frac{E}{R} \times (1 - e^{-Rt/L})$$

$$V_R = E \times (1 - e^{-Rt/L})$$
$$V_L = E\, e^{-Rt/L}$$

where

    t is the time in seconds that elapses after S is closed in position 1

    e = 2.7183 . . . and is the base of the natural logarithms

In a time of L/R seconds after S is closed in position 1, the circuit conditions are: I = 63.2% of E/R, V_R = 63.2% of E, V_L = 36.8% of E, and the rate of growth of current is 36.8% of E/L amperes per second (Fig. 6-7). In

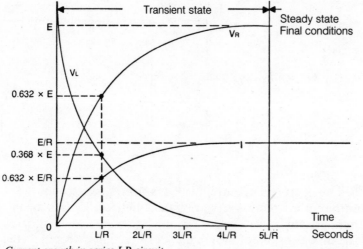

Fig. 6-7 Current growth in series LR circuit.

2 L/R seconds, the current reaches 86.5% of its final value, and after 3 L/R, and 4 L/R seconds, the corresponding percentages are 95.0% and 98.2%. When a total time of 5 L/R seconds has elapsed, the transient state is assumed to have finished and the circuit reaches its final steady-state condition in which I = E/R, V_R = E. V_L = 0, and the rate of growth of current is zero. The energy then stored in the magnetic field surrounding the inductor is $0.5\ LI^2 = 0.5\ LE^2/R^2$ joules.

If S is now switched to position 2 (Fig. 6-6), the inductor prevents any sudden change of current so that the initial conditions are I = E/R, V_R = E, V_L = −E, and the rate of current decay is E/L amperes per second.

Note that the polarity of V_L is reversed compared with the conditions when S was in position 1 because the magnetic field surrounding the inductor is now collapsing. As the current decreases, V_R and V_L fall together so that the rate of the current decay is less. The equations are

$$I = \frac{E}{R}\ e^{-Rt/L}$$

$$V_R = E\ e^{-Rt/L}$$
$$V_L = -E\ e^{-Rt/L}$$

where

t is the time in seconds that elapses after S is switched to position 2

Following a time of L/R seconds, the circuit conditions are

$$I = 36.8\%\ of\ E/R,\ V_R = 36.8\%\ of\ E$$
$$V_L = 36.8\%\ of\ E$$

The rate of current decay is 36.8% of E/L amperes per second (Fig. 6-8). In 2 L/R seconds, the current falls to 13.5% of E/R (and has therefore lost 86.5% of its original value). After 3 L/R and 4 L/R seconds, the correspond-

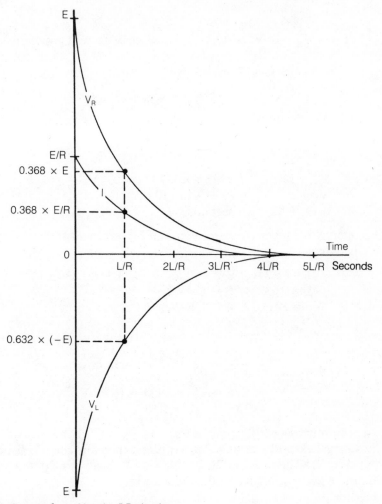

E

V_R

E/R

0.368 × E

0.368 × E/R

I

0

L/R   2L/R   3L/R   4L/R   5L/R   Seconds

Time

0.632 × (−E)

V_L

E

Fig. 6-8 Current decay in series LR circuit.

ing percentage current levels are 5.0% and 1.8%. When a total time of 5 L/R has elapsed, all quantities in the circuit are virtually zero.

Example 6-8

In Fig. 6-6, L = 26 mH, R = 3.3 kΩ, and E = 20 V. What is the time constant of this series combination? When S is closed in position 1, find the initial values of V_R, V_L and I and their subsequent values after time intervals of 5, 7.88, 20, and 40 microseconds. What is the energy stored in the inductor after the time interval of 40 microseconds?

Solution

$$\text{time constant} = L/R = \frac{26 \times 10^{-3}}{3.3 \times 10^3} \text{ seconds} = 7.88 \ \mu \text{ seconds}$$

Initial values: t = 0

I = 0, V_R = 0, V_L = 20 V and the rate of current growth is

$$\frac{E}{L} = 20/(26 \times 10^{-3})$$

$$= 769 \text{ amperes per second}$$

t = 5 microseconds

$$I = \frac{E}{R}(1 - e^{-Rt/L}) = \frac{20}{3.3}(1 - e^{-5/7.88}) = 2.85 \text{ mA}$$

$V_R = I \times R = 2.85 \text{ mA} \times 3.3 \text{ k}\Omega = 9.4 \text{ V}$
$V_L = E - V_L = 20 - 9.4 = 10.6 \text{ V}$
t = 7.88 microseconds

(This interval is equal to the time constant of the circuit.)
Therefore

$I = 63.2\%$ of E/R = 0.632 × 20/3.3 = 3.83 mA
$V_R = 0.632 \times 20 = 12.64 \text{ V}$
$V_L = 20 - 12.64 = 7.36 \text{ V}$

t = 20 microseconds

$$I = \frac{20}{3.3}(1 - e^{-20/7.88}) = 5.58 \text{ mA}$$

$$V_R = 5.58 \text{ mA} \times 3.3 \text{ k}\Omega = 18.42 \text{ V}$$

$$V_L = 20 - 18.42 = 1.58 \text{ V}$$

t = 40 microseconds
(This interval exceeds five time constants and therefore the transient state has finished.) Then: V_R = 20 V, V_L = O V and I = 20 V/3.3 kΩ = 6.06 mA. The energy stored in the inductor is

$$\frac{1}{2}LI^2 = \frac{1}{2} \times 26 \times 10^{-3} \times (6.06 \times 10^{-3})^2$$

$$= 477.5 \times 10^{-9} \text{ joules}$$
$$= 0.4775 \text{ microjoules}$$

Example 6-9

In Example 6-8, a time interval of 40 microseconds has elasped from the closing of switch S in position 1. S is now switched to position 2. What are the initial values of I, V_R, V_L, and the rate of current decay? What are the subsequent values of I, V_R, V_L after time intervals of 5, 7.88, 20, and 40 microseconds?

Solution

Initial values: t = 0.

$$I = 6.06 \text{ mA}, V_R = 20 \text{ V}, V_L = -20 \text{ V}$$

t = 5 microseconds

$$I = \frac{E}{R} e^{-Rt/L} = 6.06 \times e^{-5/7.88} = 3.21 \text{ mA}$$

$$V_R = I \times R = 3.21 \text{ mA} \times 3.3 \text{ k}\Omega = 10.6 \text{ V}$$
$$V_L = -V_R = -10.6 \text{ V}$$

t = 7.88 microseconds

$$I = 36.8\% \text{ of E/R} = 0.368 \times 6.06 \text{ mA} = 2.23 \text{ mA}$$
$$V_R = 2.23 \text{ mA} \times 3.3 \text{ k}\Omega = 7.36 \text{ V}$$
$$V_L = -7.36 \text{ V}$$

t = 20 microseconds

$$I = 6.06 \times e^{-20/7.88} = 0.48 \text{ mA}$$
$$V_R = 0.48 \text{ mA} \times 3.3 \text{ k}\Omega = 1.58 \text{ V}$$
$$V_L = -1.58 \text{ V}$$

t = 40 microseconds

$$I = 0 \text{ A}, V_R = V_L = 0 \text{ V}$$

CHAPTER SUMMARY

☐ Induced EMF $= -L \dfrac{\Delta I}{\Delta T} = -N \dfrac{\Delta \phi}{\Delta T}$ volts

☐ Inductance, $L = N \dfrac{\Delta \phi}{\Delta I}$ henrys

☐ Inductance of coil, $L = \dfrac{\mu_0 \mu_r N^2 A}{\ell}$ henrys

☐ Energy stored in magnetic field $= \frac{1}{2} LI^2$ joules

☐ Time constant of LR circuit $= \dfrac{L}{R}$ seconds

☐ Time required to reach steady-state conditions $= \dfrac{5L}{R}$ seconds

7
Electrostatics and Capacitance

WHEN STUDYING THE subjects of electrostatics and capacitance, you will learn:

□ How the charges and their separation feature in Coulomb's Law.
□ About the relationship between the charge density, the electric field intensity, and the permittivity.
□ About the electrical properties of capacitance and the definition of its unit.
□ Which physical factors determine the capacitance of a capacitor.
□ How to calculate the amount of energy stored in the electric field associated with a capacitor.
□ What occurs when a number of capacitors are connected in series.
□ What occurs when a number of capacitors are connected in parallel or series-parallel.
□ About the concept of the time constant and how it relates to the charging and discharging of a capacitor through a resistor.

In 1785, Coulomb discovered through experiments that the force between two charged bodies was directly proportional to the product of their charges and inversely proportional to the square of their separation.

COULOMB'S LAW

In equation form,

$$F = \frac{Q_1 Q_2}{4 \pi \epsilon_o d^2} = k \frac{Q_1 Q_2}{d^2} \quad \text{newtons}$$

where

$$Q, Q_2 = \text{charges measured in coulombs}$$
$$d = \text{distance between charges in meters}$$
$$k = 1/(4\,\pi\epsilon_o) = 1/(4\,\pi \times 8.85 \times 10^{-12})$$
$$= 8.99 \times 10^9 \text{ SI units}$$

The meaning of ϵ_o, the permittivity of free space, is explained later in this chapter. In the cgs system, the unit of charge is the *statcoulomb*. One coulomb = 3×10^9 statcoulombs. Between the charges there exists an electric field whose total number of lines is referred to as the electric flux. In the SI system, a single line of the electric flux is assumed to leave a positive charge of 1 coulomb and terminate on a negative charge of 1 coulomb. The number of lines is therefore equivalent to the number of coulombs and no special unit is required for the electric flux whose symbol is the Greek letter, ψ (psi).

Figure 7-1 shows two rectangular surface plates separated by a vacuum insulator of thickness d meters. For each plate, the area of one side is A square meters. The plates have been charged to a potential difference of V_C volts so that one plate carries a positive charge of Q coulombs and the other plate has an equal negative charge. The electric flux therefore consists of Q lines and the electric flux density, D, in MKS units is given by

$$D = \frac{Q}{A}$$

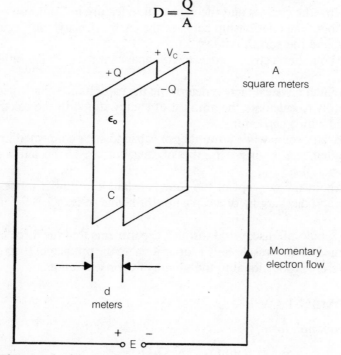

Fig. 7-1 The principle of the capacitor.

coulombs per square meter. In the cgs system, electric flux density is measured in statcoulombs per square centimeter. One coulomb per square meter $= 12 \pi \times 10^5$ statcoulombs per square centimeter.

In the region between the plates, the electric field intensity, $\mathscr{E}$, is equal to the voltage gradient, so that

$$\mathscr{E} = \frac{V_C}{d} \text{ volts per meter}$$

ELECTRIC-FIELD INTENSITY

If a charge of 1 coulomb is positioned between the plates, an electric field intensity of one volt per meter will cause a force, F, of one newton to be exerted on the charge. Therefore,

$$\mathscr{E} = \frac{F}{Q}$$

and the field intensity also can be measured in newtons per coulomb.

The cgs unit of electric field intensity is the statvolt per centimeter, where

$$1 \text{ volt per meter} = \frac{1}{3} \times 10^{-4} \text{ statvolts per centimeter}$$

The electric field intensity can be compared with the magnetic field intensity, H, which is measured in amperes per meter.

In electromagnetism, the ratio of the magnetic flux density B, in a vacuum to the magnetic field intensity, H, is the *permeability* of free space, $\mu_o = 4 \pi \times 10^{-7}$ mks units. Likewise, in electrostatics, the ratio of the electric flux density in a vacuum to the electric field intensity is called the *permittivity* of free space, whose letter symbol is ϵ_o. In equation form,

$$\frac{D}{\mathscr{E}} = \epsilon_o$$

The value of ϵ_o is 8.85×10^{-12} mks units, which could be expressed in coulombs per volt · meter. In the cgs system $\epsilon_o = 1$.

With $\mu_o = 4 \pi \times 10^{-7}$ and $\epsilon_o = 8.85 \times 10^{-12}$, it can be shown that

$$\frac{1}{\sqrt{\mu_o \epsilon_o}} = 3 \times 10^8 \text{ meters per second}$$

which is the velocity of all electromagnetic waves. One example of such a wave is light. In addition,

$$\sqrt{\frac{\mu_o}{\epsilon_o}} = 120 \pi = 377 \ \Omega$$

This is the value of the free space impedance, η_o. Clerk Maxwell discovered these relationships in 1865 and was then able to predict the existence of radio waves some twenty years before the experiments of Heinrich Hertz.

CAPACITANCE AND THE CAPACITOR

Capacitance is that property of an electrical circuit which opposes any sudden change of *voltage*. This compares with inductance, which opposes any sudden change of *current*, and resistance which merely limits the flow of current.

The property of capacitance is possessed by a capacitor, which consists of two conducting surfaces, (for example aluminum, tin foil or copper) separated by an insulator (air, mica or paper) which is referred to as the *dielectric*. Figure 7-1 therefore represents a capacitor with a vacuum as the dielectric. When a dc voltage, E, is applied across the capacitor, there is a momentary electron flow that ceases when the voltage between the plates exactly balances the applied voltage. The charge then stored depends on this voltage and some constant of the capacitor called its capacitance, C, measured in Farads (after Michael Faraday). In equation form,

$$V_C = -\frac{Q}{C}$$

The negative sign indicates that V_C balances the appled voltage, E. By Kirchhoff's Voltage Law, $E + V_C = 0$. Therefore,

$$E = \frac{Q}{C}$$

The capacitance is one farad if, when the voltage between the capacitor plates is one volt, the charge stored is one coulomb. However, the farad is too large for practical purposes so that capacitors are normally measured in either microfarads or picofarads.

$$\epsilon_o = \frac{D}{\mathscr{E}} = \frac{Q/A}{V_C/d} = \frac{Q}{V_C} \times \frac{d}{A} = C \times \frac{d}{A}$$

ϵ_o therefore can be measured in farads per meter as well as coulombs per volt $\cdot$ meter. Then

$$\text{capacitance, } C = \frac{\epsilon_o A}{d} \text{ farads}$$

Note that the capacitance is directly proportional to the area of the plates but is inversely proportional to their separation.

Because $Q = C \times V_C$, it is impossible to change the voltage between the capacitor plates without altering the value in coulombs of the charge stored. This required that a current in amperes or coulombs per second shall either charge or discharge the capacitor over a period of time. In other words, a capacitor can neither charge or discharge instantaneously, so that capacitance is that electrical property that opposes any sudden change of voltage.

RELATIVE PERMITTIVITY

If the space between the capacitor plates is filled by a particular insulating material rather than a vacuum, the capacitance increases. The ratio of a capacitor's capacitance with a certain material as the dielectric to the capacitance with a vacuum dielectric is called the material's *relative permittivity,* or dielectric constant, ϵ_r.

Table 7-1 shows the values of the dielectric constant for some commonly used insulating materials.

Table 7-1. Dielectric Constants

Material	Relative permittivity, μ_r
Air	1.0006
Alumina	4.5–8.4
Beeswax	2.66
Epoxy cast resin	3.62
Mica	5.4
Neoprene	6.60
Nylon	3.5
Porcelain	6.0–8.0
Quartz (fused)	3.75–4.1
Silica glass	3.8
Steatite	5.5–7.5
Teflon	2.0
Titanium dioxide	14–110

If the space between the capacitor plates is filled by a dielectric with a relative permittivity, ϵ_r,

$$C = \frac{\epsilon_o \epsilon_r A}{d} \text{ farads}$$

and

$$\frac{\text{electric flux density, D}}{\text{electric field intensity, } \mathscr{E}} = \epsilon_o \epsilon_r = \epsilon$$

where ϵ is the absolute permittivity

Remember that

$$\frac{\text{magnetic flux density, B}}{\text{magnetic field intensity, H}} = \mu_o \mu_r = \mu$$

where μ is the absolute permeability

Compare this equation with the one above.

Example 7-1

A capacitor consists of two sheets of tin foil, each side of which has an area of 1500 cm^2. The sheets are separated by a 0.06 mm thickness of paper dielectric whose relative permittivity is 2.25. What is the value of the capacitance?

Solution

$$\text{capacitance, } C = \frac{\epsilon_0 \epsilon_r A}{d} = \frac{8.85 \times 10^{-12} \times 2.25 \times 1500 \times 10^{-4}}{0.06 \times 10^{-3}}$$

$$= \frac{8.85 \times 2.25 \times 1.5 \times 10^{-2}}{6} \mu F$$

$$= 0.0498 \ \mu F$$

Example 7-2

The capacitor in Example 7-1 is charged to 80 V. Calculate the values of the (a) charge stored, (b) the electric flux density and (c) the electric field intensity.

Solution

(a) charge stored, $Q = C \times V_C = 0.0498 \times 10^{-6} \times 80$ C
$$= 3.98 \text{ microcoulombs}$$

(b) electric flux density, $D = \dfrac{Q}{A} = \dfrac{3.98}{1500 \times 10^{-4}}$
$$= 26.6 \text{ microcoulomb per square meter}$$

(c) electric field intensity, $= \dfrac{V_C}{d} = \dfrac{80}{0.06 \times 10^{-3}}$
$$= 1.33 \times 10^6 \text{ volts per meter}$$

Check:

$$\frac{D}{\mathscr{E}} = \frac{26.6 \times 10^{-6}}{1.333 \times 10^6} = 19.9 \times 10^{-12} \text{ farads per meter}$$

absolute permittivity, $\epsilon = \epsilon_0 \epsilon_r = 8.85 \times 10^{-12} \times 2.25$
$$= 19.9 \times 10^{-12} \text{ farads per meter}$$

ENERGY STORED IN THE ELECTRIC FIELD BETWEEN THE CAPACITOR PLATES

Let a capacitor of capacitance C farads be charged by a constant current I amperes over a period of T seconds (Fig. 7-2). The final charge, Q, is I $\times$ T coulombs and the corresponding voltage, V_C, between the capacitor plates is Q/C. Because the charging current is constant, the average voltage is $V_C/2$.

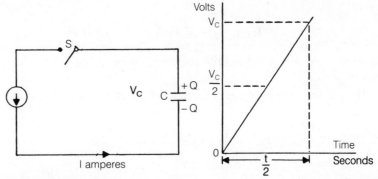

Fig. 7-2 Energy stored in a capacitor.

Energy supplied during charging = power × time

$$= I \times \frac{V_C}{2} \times T$$

$$= \frac{1}{2} V_C Q = \frac{1}{2} \frac{Q^2}{C}$$

$$= \frac{1}{2} C V_C^2 \text{ joules}$$

CAPACITORS IN SERIES

Referring to Fig. 7-3 the capacitors C_1, C_2, and C_3 are in series and are charged by the source voltage, E_T. During the charging of the capacitors, the current is the same throughout the circuit, and therefore, at the end of the charging period, each capacitor must carry the same charge, Q. Then

$$V_1 = \frac{Q}{C_1}, \quad V_2 = \frac{Q}{C_2}, \quad V_3 = \frac{Q}{C_3}$$

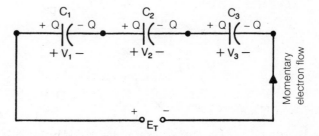

Fig. 7-3 Capacitors in series.

Also

$$E_T = V_1 + V_2 + V_3$$

$$= Q \times \left(\frac{1}{C_1} + \frac{1}{C_2} + \frac{1}{C_3} \right)$$

If C_T is the total equivalent capacitance that will store the same charge Q,

$$E_T = \frac{Q}{C_T} = Q \times \frac{1}{C_T} = V_1 \times \frac{C_1}{C_T}$$

Therefore

$$V_1 = E_T \times \frac{C_T}{C_1}$$

$$V_2 = E_T \times \frac{C_T}{C_2}$$

$$V_3 = E_T \times \frac{C_T}{C_3}$$

and

$$\frac{1}{C_T} = \frac{1}{C_1} + \frac{1}{C_2} + \frac{1}{C_3}$$

For N capacitors in series,

$$C_T = \frac{1}{\dfrac{1}{C_1} + \dfrac{1}{C_2} + \dfrac{1}{C_3} + \cdots + \dfrac{1}{C_N}}$$

If the N series capacitors all have the same value, C, then

$$C_T = \frac{C}{N}$$

For two capacitors in series,

$$C_T = \frac{C_1 C_2}{C_1 + C_2}$$

$$V_1 = E_T \times \frac{C_2}{C_1 + C_2}$$

$$V_2 = E_T \times \frac{C_1}{C_1 + C_2}$$

The formulas for capacitors in series and resistors in parallel are therefore comparable. Connecting capacitors in series reduces the capacitance, and the total capacitance is less than the value of the smallest series capacitance. Basically this is because the series arrangement effectively increases the distance between the end plates connected to the source voltage, and the capacitance is inversely proportional to this distance. Because a capacitor has a dc working-voltage rating, you can use the series arrangement to distribute the source voltage between the capacitors so that the voltage across an individual capacitor does not exceed its rating.

CAPACITORS IN PARALLEL

In Fig. 7-4, the capacitors C_1, C_2, C_3 are in parallel, and each is charged by the source voltage E_T. Therefore: $Q_1 = C_1 E_T$, $Q_2 = C_2 E_T$, and $Q_3 = C_3 E_T$.

The total charge, Q_T is the sum of the individual charges stored in the capacitors. In equation form, $Q_T = Q_1 + Q_2 + Q_3 = E_T \times (C_1 + C_2 + C_3)$.

If C_T is the total capacitance, $Q_T = E_T C_T$.

Therefore $C_T = C_1 + C_2 + C_3$.

For N capacitors in parallel, $C_1 = C_1 + C_2 + C_3 + \cdots + C_N$.

If N parallel capacitors all have the same value, C, then $C_T = NC$.

The total equivalent capacitance of capacitors in parallel is the sum of their individual capacitances. The parallel arrangement increases the effective surface area and therefore the capacitance, which is directly proportional to A.

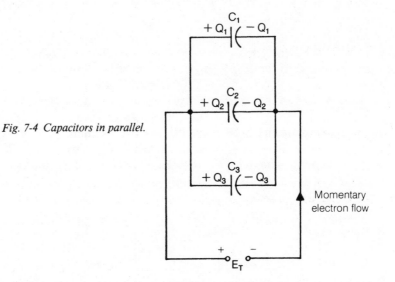

Fig. 7-4 Capacitors in parallel.

If a capacitor is made up of N parallel plates with alternate plates connected as in Fig. 7-5, the total capacitance is

$$C_T = \frac{\epsilon_o \epsilon_r (N-1) A}{d}$$

where ϵ_o, ϵ_r, A and d are defined as before.

Example 7-3

In Fig. 7-3, $C_1 = 10\ \mu F$, $C_2 = 16\ \mu F$, $C_3 = 8\ \mu F$, and $E_T = 56$ V. Find the values of C_T, V_1, V_2, and V_3. What is the amount of charge and the energy stored in each capacitor?

Solution

$$\text{total capacitance, } C_T = \frac{1}{\dfrac{1}{10} + \dfrac{1}{16} + \dfrac{1}{8}}$$

$$= 3.478\ \mu F$$

$$\text{Charge stored in each capacitor, } Q = 3.478 \times 10^{-6}\ F \times 56\ V$$
$$= 194.8\ \mu C.$$

Then

$$V_1 = \frac{194.8\ \mu C}{10\ \mu F} = 19.48\ V$$

$$V_2 = \frac{194.8\ \mu C}{16\ \mu F} = 12.17\ V$$

$$V_3 = \frac{194.8\ \mu C}{8\ \mu F} = 24.35\ V$$

Check:

$$E = V_1 + V_2 + V_3 = 19.48 + 12.17 + 24.35 = 56\ V$$
$$\text{energy stored in the 10-}\mu F \text{ capacitor} = \tfrac{1}{2} \times 194.8\ \mu C \times 19.48\ V$$
$$= 1897\ \mu \text{ joules}$$
$$\text{energy stored in the 16-}\mu F \text{ capacitor} = \tfrac{1}{2} \times 194.8\ \mu C \times 12.17\ V$$
$$= 1185\ \mu \text{ joules}$$
$$\text{energy stored in the 8-}\mu F \text{ capacitor} = \tfrac{1}{2} \times 194.8\ \mu C \times 24.35\ V$$
$$= 2372\ \mu \text{ joules}$$
$$\text{total energy stored} = 1897 + 1185 + 2372 = 5454\ \mu \text{ joules}$$

Check:

$$\text{total energy stored} = \frac{1}{2} \times 194.8\ \mu C \times 56\ V = 5454\ \mu \text{ joules}$$

Example 7-4

In Fig. 7-4, $C_1 = 10\ \mu F$, $C_2 = 16\ \mu F$, $C_3 = 8\ \mu F$, and $E_T = 56$ V. Find the values of C_T, Q_T, Q_1, Q_2, and Q_3. What is the energy stored in each capacitor?

Solution

$$\text{total capacitance, } C_T = 10 + 16 + 8 = 34\ \mu F$$
$$\text{total charge stored, } Q_T = 34\ \mu F \times 56\ V = 1904\ \mu C$$

Then

$$Q_1 = 10 \ \mu F \times 56 \ V = 560 \ \mu C$$
$$Q_2 = 16 \ \mu F \times 56 \ V = 896 \ \mu C$$
$$Q_3 = 8 \ \mu F \times 56 \ V = 448 \ \mu C$$

Check

$$\text{total charge stored, } Q_T = Q_1 + Q_2 + Q_3$$
$$= 560 + 896 + 448$$
$$= 1904 \ \mu C$$

Energies stored in C_1, C_2, C_3 are respectively:

½ × 560 × 56, ½ × 896 × 56, ½ × 448 × 56,

or 15680, 25088, 12544 μ joules.
total energy stored is 15680 + 25088 + 12544 = 53312 μ joules.

Check:

$$\text{total energy stored is } \frac{1}{2} \times Q_T \times E_T = \frac{1}{2} \times 1904 \ \mu C \times 56 \ V$$
$$= 53312 \ \mu \text{ joules}$$

Example 7-5

A 40-μF capacitor is charged from a 200 V source. Another 20-μF capacitor is charged from a 100 V source. Immediately after the capacitors are disconnected, they are correctly paralleled. What is (a) the voltage across the combination and (b) the total energy before and after the parallel connection?

Solution

(a) charge stored in the 40-μF capacitor = 200 V × 40 μF
$$= 8000 \ \mu C$$
charge stored in the 20 μF capacitor = 100 V × 20 μF
$$= 2000 \ \mu C$$
total charge = 8000 + 2000 = 10000 μC
total capacitance of the parallel combination = 40 μF + 20 μF
$$= 60 \ \mu F$$
voltage across the parallel combination = $\dfrac{10000}{60}$ = 166.7 V

(b) energy stored in the 40 μF capacitor = $\dfrac{1}{2}$ × 8000 μC × 200 V

$$= 0.8 \text{ joule}$$

energy stored in the 20 μF capacitor = $\dfrac{1}{2}$ × 2000 μC × 100 V

$$= 0.1 \text{ joule}$$

total energy stored before the parallel connection $= 0.8 + 0.1$
$$= 0.9 \text{ joule}$$

total energy stored after the parallel connection $= \dfrac{1}{2} \times 10000 \text{ C} \times$
$166.7 \text{ V} = 0.833 \text{ joule}$

Note that energy is lost as the result of the parallel connection. If you neglect the resistance of any connecting wires, the lost energy appears in the form of the spark, which appears when the parallel connection is made.

Example 7-6

A capacitor is made with nine conducting surfaces with alternate surfaces connected as shown in Fig. 7-5. The area of one side of one surface is 150 cm², and the surfaces are separated by sheets of mica with a thickness of 0.4 mm and a relative permittivity of 5. What is the value of the capacitance?

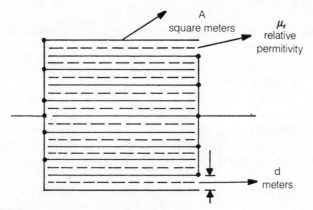

Fig. 7-5 Multi-plate capacitor.

Solution

$N = 9$, $A = 150 \times 10^{-4}$ m², $d = 0.4 \times 10^{-3}$m, $\epsilon_r = 5$.

Then

$$C = \frac{\epsilon_o \epsilon_r (N-1)A}{d} = \frac{8.85 \times 10^{-12} \times 5 \times 8 \times 150 \times 10^{-4}}{0.4 \times 10^{-3}} \text{ F}$$

$$= 8.85 \times 10 \times 150 \text{ pF}$$
$$= 0.0133 \ \mu\text{F}$$

TIME CONSTANT OF A CAPACITOR AND A RESISTOR IN SERIES

In the circuit of Fig. 7-6, the capacitor, C, is uncharged. The circuit conditions immediately after the switch S is closed in position 1, are

$$Q = 0, \ V_C = 0, \ V_R = E$$

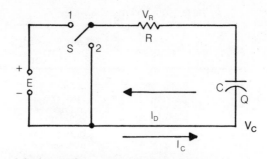

Fig. 7-6 Charge and discharge of a capacitor through a resistor.

and the charging current is

$$I_C = \frac{E}{R} \text{ (maximum value)}$$

Q is the charge, measured in coulombs, that initially must be zero because a capacitor can neither charge nor discharge instantaneously. Because no voltage can appear across the capacitor until time has elapsed after the switch is closed in position 1, the capacitor initially must behave as a *short circuit*.

The time constant of the circuit is CR seconds, where C is measured in farads and R in ohms. After a time interval of CR seconds from the instant S is closed in position 1, the circuit conditions as shown in Fig. 7-7 are

$$Q = 63.2\% \text{ of CE coulombs}$$
$$V_C = 63.2\% \text{ of E}$$
$$V_R = 36.8\% \text{ of E}$$
$$I_C = 36.8\% \text{ of } \frac{E}{R}$$

You can find the order of the time constant value from Table 7-2.

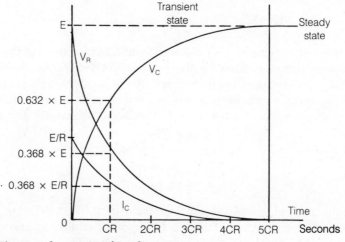

Fig. 7-7 Charging of a capacitor through a resistor.

Table 7-2. CR Time Constants

C	R	Time Constant CR
farads	ohms	seconds
μF	ohms	microseconds
μF	kΩ	milliseconds
μF	MΩ	seconds
pF	ohms	picoseconds
pF	kΩ	nanoseconds
pF	MΩ	microseconds

Following an interval of 2 CR seconds from the instant S is closed in position 1, the capacitor has acquired 86.5% of its final charge, CE coulombs, (and the voltage across the capacitor is 0.865 E). After 3 CR and 4 CR seconds, the charge percentages are respectively 95.0% and 98.2%. After 5 CR seconds, the circuit is assumed to have reached its steady state conditions, which are $Q = CE$, $V_C = E$, $V_R = 0$, $I_C = 0$. The fully charged capacitor then behaves as an *open* circuit. The following equations refer to the charging of the capacitor during the transient state.

$$Q = CE(1 - e^{-t/CR})$$
$$V_C = E(1 - e^{-t/CR})$$

$$I_C = \frac{E}{R} e^{-t/CR}$$

$$V_R = E\, e^{-t/CR}$$

where

$e = 2.7183$ and is the base of the natural logarithms

If S is now switched to position 2, the initial conditions are

$$Q = CE,\ V_C = E,\ V_R = -E,\ I_D = -E/R$$

The negative signs for the values of V_R and I_D indicate that the polarity of the voltage drop across the resistor, V_R, has reversed because the discharge current, I_D, is in the opposite direction to the previous charging current, I_C; I_D and V_R therefore are shown below the time axis in Fig. 7-8. After an interval equal to one time constant (CR seconds), the conditions are

$$Q = 36.8\% \text{ of CE}$$
$$V_C = 36.8\% \text{ of E}$$
$$V_R = 36.8\% \text{ of } (-E)$$
$$I_D = 36.8\% \text{ of } (-E/R)$$

Following an interval of 2 CR from the instant S is switched to position 2, the capacitor's charge has fallen to 13.5% of CE coulombs (the capacitor therefore has lost 86.5% of its initial charge). After 3 CR seconds, the per-

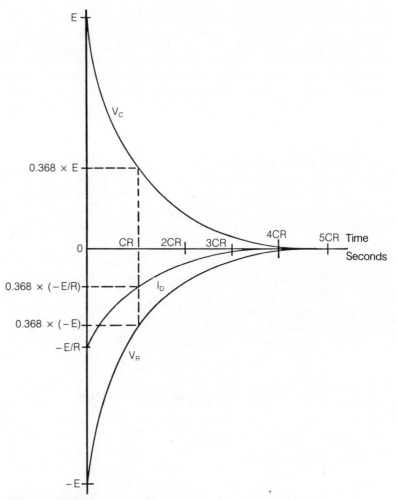

Fig. 7-8 Discharge of a capacitor through a resistor.

centages for the capacitor's remaining charge are respectively 5.0% and 1.8%. The equations for the transient discharge state are

$$Q = CE \, e^{-t/CR}$$

$$V_C = E \, e^{-t/CR}$$

$$I_D = -\frac{E}{R} \, e^{-t/CR}$$

$$V_R = -E \, e^{-t/CR}$$

In a total time of 5 CR seconds, the capacitor is assumed to have fully discharged and all circuit quantities are zero.

Example 7-7

In Fig. 7-6, C = 50 pFs, R = 1 MΩ, and E = 80 V. What is the time constant of the circuit? If S is closed in position 1, what are the values of Q, V_C, V_R, and I_C after time intervals of 0, 30, 50, 90, 180, and 300 μ seconds? What is the energy stored in the capacitor after 250 μ seconds?

Solution

time constant = C × R = 50 μ seconds
initial conditions (t = 0)

$$V_C = 0 \text{ V}, \ V_R = 80 \text{ V}, \ Q = 0 \text{ C}, \ I_C = \frac{80 \text{ V}}{1 \text{ M}\Omega} = 80 \ \mu\text{A}$$

time = 30 μ seconds

$$V_C = E(1 - e^{-t/CR}) = 80(1 - e^{-30/50}) = 36.1 \text{ V}$$
$$V_R = E - V_C = 80 - 36.1 = 43.9 \text{ V}$$
$$Q = C \times V_C = 50 \text{ pF} \times 36.1 \text{ V} = 1805 \text{ pC}$$

$$I_C = \frac{V_R}{R} = \frac{43.9 \text{ V}}{1 \text{ M}\Omega} = 43.9 \ \mu\text{A}$$

time = 50 μ seconds (one time constant)

$$V_C = 80 \times 0.632 = 50.6 \text{ V}$$
$$V_R = 80 - 50.6 = 29.4 \text{ V}$$
$$Q = 50 \text{ pF} \times 50.6 \text{ V} = 2530 \text{ pC}$$

$$I_C = \frac{29.4 \text{ V}}{1 \text{ M}\Omega} = 29.4 \ \mu\text{A}$$

time = 90 μ seconds

$$V_C = 80(1 - e^{-90/50}) = 66.8 \text{ V}$$
$$V_R = 13.2 \text{ V}$$
$$Q = 50 \text{ pF} \times 66.8 \text{ V} = 3340 \text{ pC}$$

$$I_C = \frac{13.2 \text{ V}}{1 \text{ M}\Omega} = 13.2 \ \mu\text{A}$$

time = 180 μ seconds

$$V_C = 80(1 - e^{-180/50}) = 77.8 \text{ V}$$
$$V_R = 2.2 \text{ V}$$
$$Q = 50 \text{ pF} \times 77.8 \text{ V} = 3890 \text{ pC}$$

$$I_C = \frac{2.2 \text{ V}}{1 \text{ M}\Omega} = 2.1 \ \mu\text{A}$$

time = 300 μ seconds (six time constants)

The circuit is now in its steady state condition.

$$V_C = 80 \text{ V}$$

$$V_R = 0 \text{ V}$$
$$Q = 50 \text{ pF} \times 80 \text{ V} = 4000 \text{ pC}$$
$$I_C = 0 \text{ A}$$

Energy stored in the capacitor after 250 μ seconds (five time constants) is

$$\frac{1}{2} CE^2 = \frac{1}{2} \times 50 \times 10^{-12} \times 80^2$$
$$= 1.6 \times 10^{-7} \text{ joule}$$

Example 7-8

In Fig. 7-6, S is closed in position 1, and the circuit has reached its steady-state condition. S is now switched to position 2. From that instant, what are the values of Q, V_C, V_R, and I_D after time intervals of 0, 30, 50, 90, 180, and 300 μ seconds?

Solution

initial conditions (t = 0)

$V_C = 80$ V, $V_R = -80$ V, Q = 50 pFs $\times$ 80 V = 4000 pC, and I_D
  $= -80$ V/1 MΩ = -80 μA

time = 30 μ seconds

$$V_C = E \, e^{-t/CR} = 80 \times e^{-30/50} = 43.9 \text{ V}$$
$$V_R = -43.9 \text{ V}$$
$$Q = 50 \text{ pF} \times 43.9 \text{ V} = 2195 \text{ }\mu\text{C}$$
$$I_D = -43.9 \text{ V}/1 \text{ M}\Omega = -43.9 \text{ }\mu\text{A}$$

time = 50 μ seconds (one time constant)

$$V_C = 80 \times 0.368 = 29.4 \text{ V}$$
$$V_R = -29.4 \text{ V}$$
$$Q = 50 \text{ pF} \times 29.4 \text{ V} = 1470 \text{ pC}$$
$$I_D = -29.4 \text{ V}/1 \text{ M}\Omega = -29.4 \text{ }\mu\text{A}$$

time = 90 μ seconds

$$V_C = 80 \times e^{-90/50} = -13.2 \text{ V}$$
$$V_R = -13.2 \text{ V}$$
$$Q = 50 \text{ pF} \times 13.2 \text{ V} = 661 \text{ pC}$$
$$I_D = -13.2 \text{ V}/1 \text{ M}\Omega = -13.2 \text{ }\mu\text{A}$$

time = 180 μ seconds

$$V_C = 80 \times e^{-180/50} = 2.2 \text{ V}$$
$$V_R = -2.2 \text{ V}$$
$$Q = 50 \text{ pF} \times 2.2 \text{ V} = 110 \text{ pC}$$
$$I_D = -2.2 \text{ V}/1 \text{ M}\Omega = -2.2 \text{ }\mu\text{A}$$

time = 300 μ seconds (six time constants)

The capacitor is now completely discharged. $Q = 0$ C, $V_C = V_R = 0$ V, $I_D = 0$ A.

Example 7-9

In Fig. 7-9, the capacitor initially is uncharged, and the switch S is then closed. What are the values of I_1, I_2, I_3, and the potential at X at the start and finish of the transient state? What is the time constant of the circuit, and what is the energy stored in the capacitor at the end of the transient state?

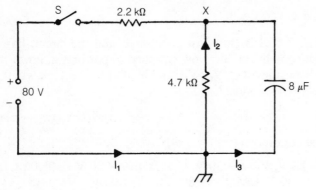

Fig. 7-9 Circuit for example 7-9.

Solution

Because the capacitor cannot charge instantaneously, the initial potential at X is *zero* volts and $I_2 = 0$ A. Therefore $I_1 = I_3 = 80$ V/2.2 kΩ = 36.4 mA. At the end of the transient state, the capacitor is fully charged and $I_3 = 0$ A. Then $I_1 = I_2 = 80$ V/(2.2 kΩ + 4.7 kΩ) = 11.6 mA, and the potential at X is 11.6 mA $\times$ 4.7 kΩ = 54.5 V. This is also the voltage to which the capacitor is charged so that the final energy stored is $\frac{1}{2} \times 8\ \mu$F $\times$ (54.5 V)^2 = 11890 μ joules.

With respect to the capacitor the resistors are in parallel (this may be shown by the use of Thévenin's Theorem in chapter 8. The circuit's time constant is

$$\frac{2.2 \times 4.7}{2.2 + 4.7} \text{ k}\Omega \times 8\ \mu\text{F} = 12 \text{ milliseconds}$$

rounded off

Example 7-10

In Fig. 7-10, the capacitors are uncharged initially and the switch S is then closed. What are the values of I and the potentials at X, Y, and Z at the start and at the finish of the transient state? What is the value of the circuit's time constant?

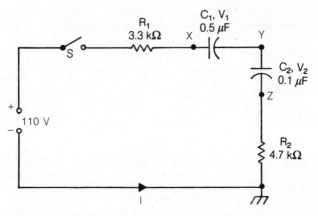

Fig. 7-10 Circuit for example 7-10.

Solution

Because the capacitors cannot charge instantaneously, the supply voltage of 110 V initially must be divided between the 3.3 kΩ and 4.7 kΩ resistors. Therefore I is 110 V/(3.3 kΩ + 4.7 kΩ) = 13.75 mA, and the potentials at X, Y, and Z are all +13.75 mA $\times$ 4.7 kΩ = +64.6 V.

At the end of the transient state, the capacitors are charged fully, and therefore I must be zero. The 110 V now must be divided between the capacitors so that

$$V_2 = E_T \times \frac{C_1}{C_1 + C_2} = 110 \text{ V} \times \frac{0.5 \ \mu F}{0.5 \ \mu F + 0.1 \ \mu F} = 91.7 \text{ V}$$

The potentials at X, Y, and Z are respectively +110 V, +91.7 V, and 0 V.

The total capacitance is

$$\frac{0.5 \times 0.1}{0.5 + 0.1} = 0.0833 \ \mu F$$

so that the circuit's time constant is (3.3 + 4.7) kΩ $\times$ 0.0833 μF = 0.66 millisecond.

CHAPTER SUMMARY

☐ Q (coulombs) = C (farads) $\times$ V_C (volts)

☐ Electric field intensity, $\mathscr{E} = \dfrac{V_C}{d}$ volts per meter or newtons per coulomb

☐ Electric flux density, $D = \dfrac{D}{A}$ coulombs per square centimeter

☐ Absolute permittivity, $\epsilon = \dfrac{D}{\mathscr{E}} = \epsilon_0 \epsilon_r$ farads per meter

Table 7-3. Converting SI/cgs Units

Quantity	SI Units	cgs Electrostatic Units
Charge, Q	1 coulomb	3×10^9 statcoulombs
EMF, E, and P.D.V	1 volt	1/300 statvolt
Current, I	1 ampere	3×10^9 statamperes
Resistance, R	1 ohm	1.11×10^{-12} statohms
Capacitance, C	1 farad	9×10^{11} statfards
Electric field intensity, $\mathscr{E}$	1 volt/meter 1 newton/coulomb	$\frac{1}{3} \times 10^{-4}$ statvolt per centimeter
Electric flux density, D	1 coulomb per square meter	$12 \pi \times 10^5$ statcoulombs per square centimeter
Permittivity of free space, ϵ_o	8.85×10^{-12} farads per meter	1

☐ Relative permittivity or dielectric constant of an insulating material.

$$\epsilon_r = \frac{\text{capacitance with insulating material as the dielectric}}{\text{capacitance with a vacuum dielectric}}$$

☐ Capacitance, $C = \dfrac{\epsilon_o \epsilon_r}{d} = \dfrac{\epsilon A}{d}$ farads

☐ Energy stored in a capacitor $= \dfrac{1}{2} CV_C^2 = \dfrac{1}{2} QV_C = \dfrac{1}{2} \dfrac{Q^2}{C}$ joules

☐ Capacitors in series: $\dfrac{1}{C_T} = \dfrac{1}{C_1} + \dfrac{1}{C_2} + \dfrac{1}{C_3} + \cdots \dfrac{1}{C_N}$

$$V_1 = E_T \times \frac{C_T}{C_1}, \; V_2 = E_T \times \frac{C_T}{C_2}, \text{ etc.}$$

☐ Two capacitors in series:

$$C_T = \frac{C_1 C_2}{C_1 + C_2}, \; V_1 = E_T \times \frac{C_2}{C_1 + C_2}, \; V_2 = E_T \times \frac{C_1}{C_1 + C_2}$$

☐ Capacitors in parallel:

$$C_T = C_1 + C_2 + C_3 \cdots + C_N$$
$$Q_T = Q_1 + Q_2 + Q_3 \cdots + Q_N$$

☐ Multiplate capacitor:

$$C = \frac{\epsilon_o \epsilon_r (N - 1) A}{d}$$

☐ Charge and discharge of capacitor through resistor:
Time constant in seconds = C (farads) $\times$ R (ohms). Total time of transient state = 5 CR seconds.

8
Network Theorems

BY STUDYING THESE aids to circuit analysis, you will learn:

☐ How to employ Kirchhoff's Laws in order to analyze complex networks and circuits with more than one voltage source.

☐ To apply mesh analysis in order to solve complex networks as well as circuits with more than one voltage source.

☐ The superposition theorem and its application to the solution of circuits with more than one electrical source.

☐ How Thévenin's theorem can reduce a complex network to a single constant voltage source.

☐ How Norton's theorem can reduce a complex network to a single constant-current source.

☐ To employ nodal analysis in the solution of circuits that contain constant-current sources.

☐ How delta → wye and wye → delta transformations can sometimes be used to simplify complex circuits.

A network can be defined as any electrical arrangement containing resistors and sources. A single network can consist of a single closed circuit (or mesh) while more complex networks could contain a number of meshes that are interdependent.

You can find the current through, and the voltage across, any resistor of a network by applying Ohm's and Kirchhoff's Laws, but in the case of a complex network, the process is lengthy and tedious because of the need to solve a large number of simultaneous equations. A series of network theorems therefore have been formulated to simplify the calculations.

Some of the theorems in this chapter have universal application while others are restricted to circuits containing linear resistances. A *linear resistance* is defined as any resistance that obeys Ohm's Law so that the voltage across the resistance is directly proportional to its current. Resistors fall into this category while semiconductor devices and tubes do not.

In this chapter, the theorems are used only with dc circuits, although they are equally valid for ac circuits. However before you can apply the theorems to general ac networks, it is necessary to have a background in complex algebra (chapter 15).

I state the following theorems without proof, but I include examples to illustrate their application.

SUPERPOSITION THEOREM

If a network of linear resistances contains more than one source, the current flowing at any point is the algebraic sum of the currents that would flow at that point if each source was considered separately while at the same time all other sources were replaced by their equivalent internal resistances. This last step is achieved by short-circuiting all sources of constant voltage and open-circuiting all sources of constant current.

Example 8-1

In the circuit of Fig. 8-1, calculate the value of the load current, I_L.

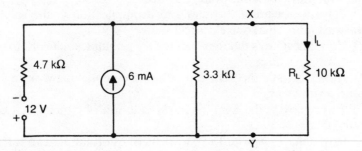

Fig. 8-1 Circuit illustrating various network theorems.

Solution

The problem may be solved in a number of ways.

METHOD 1: Millman's Theorem (see chapter 4) Convert the 12 V source into the equivalent current generator (Fig. 8-2). The constant current is

$$\frac{12 \text{ V}}{4.7 \text{ k}\Omega} = 2.55 \text{ mA}$$

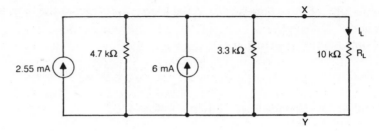

Fig. 8-2 Solution by Millman's theorem.

so that the total generator current is 6 mA + 2.55 mA = 8.55 mA. The total equivalent internal resistance is

$$\frac{4.7 \times 3.3}{4.7 + 3.3} = 1.94 \text{ k}\Omega$$

Then by the current division rule

$$\text{load current, } I_L = 8.55 \text{ mA} \times \frac{1.94 \text{ k}\Omega}{10 \text{ k}\Omega + 1.94 \text{ k}\Omega} = 1.39 \text{ mA}$$

METHOD 2: Kirchhoff's Voltage and Current Laws Convert the 6 mA constant current generator into its constant-voltage equivalent. The constant voltage is 6 mA × 3.3 kΩ = 19.8 V that is in series with the 3.3 kΩ resistor. Now you must specify the current flowing in the various parts of the circuit and insert the polarities of the voltage drops across the resistor (Fig. 8-3). In most cases, the direction of the electron flow is obvious, but sometimes the direction is unclear. This does not present any difficulty because if you choose the wrong direction, your analysis will show that the value of the current is negative. To write down the equation for a loop using Kirchhoff's Voltage Law, the convention is to move clockwise round the loop (starting at any point) and to regard the voltage as positive if you encounter negative polarity first. Kirchhoff's Voltage Law states that the

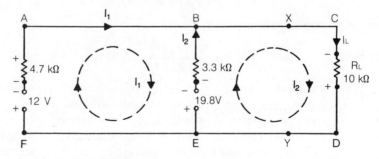

Fig. 8-3 Solution by Kirchhoff's laws.

algebraic sum of the voltages around the loop is zero so that the loop equations are

Loop ABEFA (starting at A):

$$-3.3 I_2 + 19.8 - 12 + 4.7 I_1 = 0$$
$$4.7 I_1 - 3.3 I_2 = -7.8$$

The unit of the currents is the milliampere.

Loop EBCDE (starting at E):

$$-19.8 + 3.3 I_2 + 10 I_L = 0.$$
$$3.3 I_2 + 10 I_L = 19.8$$

Loop CDEFABC (starting at C):

$$10 I_L - 12 + 4.7 I_1 = 0$$
$$4.7 I_1 + 10 I_L = 12$$

The third loop equation does not provide additional information because it is obtained by adding the two preceding equations. Therefore, any two of the three loop equations are sufficient for the analysis.

In using Kirchhoff's Current Law, the convention is to regard currents entering a particular point as positive. Therefore

$$I_1 + I_2 - I_L = 0$$

or

$$I_2 = I_L - I_1$$

Substituting for I_L in the equation for loop EBCDE,

$$3.3(I_L - I_1) + 10 I_L = 19.8$$
$$-3.3 I_1 + 13.3 I_L = 19.8$$

Multiplying the equation for loop CDEFABC by 3.3 and the preceding equation by 4.7,

$$15.51 I_1 + 33 I_L = 39.6$$
$$-15.51 I_1 + 62.51 I_L = 93.06$$

Adding these two equations,

$$\text{load current, } I_L = \frac{132.66}{95.51} = 1.39 \text{ mA}$$

METHOD 3: Mesh Current Analysis   In this method, the convention is to ascribe clockwise mesh currents i_1, i_2 (electron flow) to two of the loops. In contrast to an analysis by Kirchhoff's laws, an individual component can carry one or more mesh currents. In Fig. 8-3, the total current through the 3.3 kΩ resistance is the algebraic sum of i_1 and i_2. The mesh equations will be

$$4.7 i_1 + 3.3 (i_1 - i_2) + 19.8 - 12 = 0$$
$$8 i_1 - 3.3 i_2 = -7.8$$

and

$$3.3\,(i_2 - i_i) + 10\,i_2 - 19.8 = 0$$
$$-3.3\,i_1 + 13.3\,i_2 = 19.8$$

Multiplying these two equations by 3.3 and 8 respectively,

$$26.4\,i_1 - 10.89\,i_2 = -25.74$$
$$-26.4\,i_1 + 106.4\,i_2 = 158.4$$

Then:

$$(106.4 - 10.89)\,i_2 = 158.4 - 25.74$$
$$95.51\,i_2 = 132.66$$

$$\text{load current,} \quad i_2 = \frac{132.66}{95.51} = 1.39 \text{ mA}$$

METHOD 4: Superposition Theorem

Step 1   Short out the 12 V constant EMF (Fig. 8-4A). By the current-division rule, the value of I_{L1} of the load current from the 6 mA generator is 6 mA $\times R_T/10$ kΩ where R_T is the total equivalent resistance of 3.3 kΩ, 4.7 kΩ, and 10 kΩ in parallel.

Therefore

$$R_T = \frac{1}{\dfrac{1}{3.3} + \dfrac{1}{4.7} + \dfrac{1}{10}}$$

$$= \frac{1}{0.303 + 0.213 + 0.1}$$

$$= \frac{1}{0.616} = 1.62 \text{ k}\Omega$$

and

$$I_{L1} = 6 \text{ mA} \times 1.62 \text{ k}\Omega/10 \text{ k}\Omega = 0.972 \text{ mA}$$

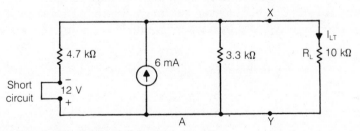

Fig. 8-4 A Solution by superposition theorem, step one.

Step 2   Open-circuit (remove) the 6 mA current generator (Fig. 8-4B). The total resistance of the remaining circuit is

$$4.7 \text{ k}\Omega + \frac{3.3 \text{ k}\Omega \times 10 \text{ k}\Omega}{3.3 \text{ k}\Omega + 10 \text{ k}\Omega} = 7.10 \text{ k}\Omega.$$

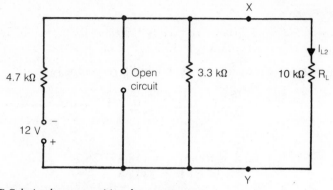

Fig. 8-4 B Solution by superposition theorem, step two.

The load current due to the 12 V source is:

$$I_{L2} = \frac{12 \text{ V}}{7.18 \text{ k}\Omega} \times \frac{3.3 \text{ k}\Omega}{10 \text{ k}\Omega + 3.3 \text{ k}\Omega} = 0.415 \text{ mA}$$

Because I_{L1} and I_{L2} flow in the same direction, the total load current by superposition is $0.974 + 0.415 = 1.39$ mA, rounded off.

THÉVENIN'S THEOREM

The current in a load resistance connected between two terminals X, Y of a network of resistances and generators is the same as if this load resistance were connected to a simple constant-voltage generator whose EMF is the open-circuit voltage measured between X and Y and whose internal resistance is the resistance of the network looking back into the terminals X, Y with all generators replaced by resistances equal to their internal resistances. You accomplish this last step by short-circuiting all sources of constant voltage and open-circuiting all sources of constant current.

Figure 8-5 is used to illustrate Thévenin's Theorem. The Thévenin voltage, E_{TH}, is the open-circuit voltage between the terminals X, Y with R_L disconnected, R_{Th} is the Thévenin resistance measured between X and Y with all generators replaced by resistances equal to their internal resistances.

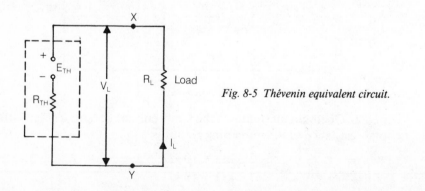

Fig. 8-5 Thévenin equivalent circuit.

Example 8-2

In the circuit of Fig. 8-1, calculate the value of I_L, using Thévenin's Theorem.

Solution

Step 1 Remove the load R_L and convert the constant-current generator into its constant-voltage equivalent with an EMF equal to 6 mA $\times$ 3.3 kΩ = 19.8 V (Fig. 8-3).

The current, I, is

$$\frac{19.8 \text{ V} - 12 \text{ V}}{4.7 \text{ k}\Omega + 3.3 \text{ k}\Omega} = \frac{7.8}{8} = 0.975 \text{ mA}$$

Then

$$E_{OC} = E_{TH} = 19.8 \text{ V} - (3.3 \text{ k}\Omega \times 0.975 \text{ mA}) = 16.58 \text{ V}$$

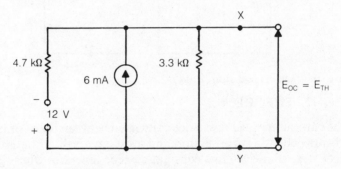

Fig. 8-6 A Solution by Thévenin's theorem, step one.

Step 2 Short-circuit the 12 V constant-voltage source and open-circuit the 6 mA constant-current source so that

$$R_{XY} = \frac{4.7 \times 3.3}{4.7 + 3.3} = 1.94 \text{ k}\Omega \text{ (Fig. 8-6B)}$$

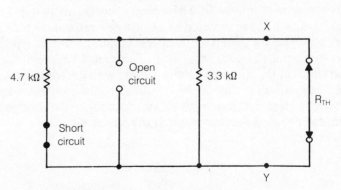

Fig. 8-6 B Solution by Thévenin's theorem, step two.

Then the Thévenin equivalent circuit is as shown in Fig. 8-6C.

Step 3   Reconnect the load R_L between X and Y. The load current is

$$I_L = \frac{16.58 \text{ V}}{1.94 \text{ k}\Omega + 10 \text{ k}\Omega} = 1.39 \text{ mA}$$

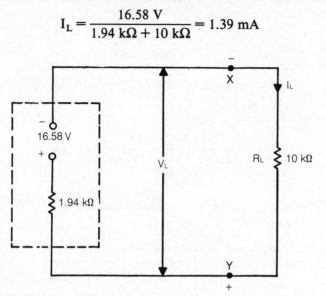

Fig. 8-6 C The Thévenin equivalent circuit.

NORTON'S THEOREM

The current in a load resistance connected between two terminals X, Y of a network consisting of generators and resistances is the same as if the load resistance were connected to a constant-current generator whose generated current, I_N, is equal to the short-circuit current measured between X and Y. This constant-current generator has infinite internal resistance but is placed in parallel with a resistance, R_N, equal to the resistance of the network looking back into the terminals X and Y with all generators replaced by resistances equal to their internal resistances (Fig. 8-7). You accomplish this last step by short-circuiting all sources of constant voltage and open-circuiting all sources of constant current.

This theorem is similar to Thévenin's theorem in that it enables a complicated network to be replaced by a single generator and a resistance. In this case, however, the generator is of the constant-current type and the resistance is in parallel, while in the case of Thévenin's Theorem, the equivalent generator is of the constant-voltage variety with the resistance in series. Because the procedure for finding R_{XY} is the same for both theorems, $R_N = R_{TH}$. The equivalent circuits effectively are identical to one another and can be transposed by using the results found in chapter 4.

Therefore:

$$I_N = \frac{E_{TH}}{R_{TH}}$$

$$E_{TH} = I_N \times R_N$$

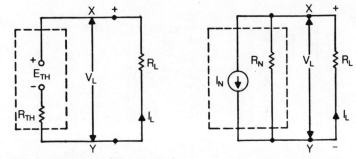

Fig. 8-7 Equivalent Thévenin and Norton circuits.

and

$$R_N = R_{TH}$$

Example 8-3

In the circuit of Fig. 8-1, calculate the value of I_L by using Norton's Theorem.

Solution

Step 1 Remove the load R_L and replace it by a short-circuit. Then convert the constant-voltage generator into its constant-current equivalent with a value of $12 \text{ V}/4.7 \text{ k}\Omega = 2.55 \text{ mA}$, (Fig. 8-8A). Then $I_N = 2.55 \text{ mA} + 6 \text{ mA} = 8.55 \text{ mA}$.

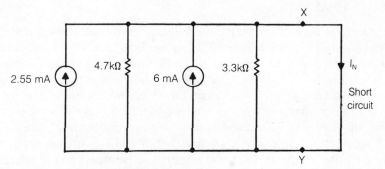

Fig. 8-8 A Solution by Norton's theorem, step one.

Step 2 Open-circuit the constant-current sources (Fig. 8-8B). The $R_{XY} = R_N = 1.94 \text{ k}\Omega$, which is equal in value to the series Thévenin resistance. The equivalent Norton generator appears in Fig. 8-8C.

Step 3 Remove the short circuit between X and Y and replace the load. If you use the constant-division rule in the Norton equivalent circuit, the load current is

$$I_L = 8.55 \text{ mA} \times \frac{1.94 \text{ k}\Omega}{1.94 \text{ k}\Omega + 10 \text{ k}\Omega} = 1.39 \text{ mA}$$

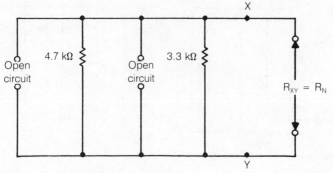

Fig. 8-8 B Solution by Norton's theorem, step two.

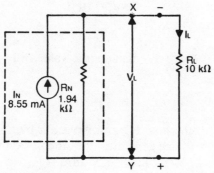

Fig. 8-8 C The Norton equivalent circuit.

NODAL ANALYSIS

A *node* is a junction point in an electrical circuit. This method of analysis consists of using Kirchhoff's Current Law to obtain the equations for a particular node. Its application can be illustrated by again using the example of Fig. 8-1. The formal method of nodal analysis requires that only current sources are included; the circuit therefore has been redrawn as in Fig. 8-9 with the constant-voltage generator replaced by its constant-current equivalent. A ground is included to provide a convenient reference node.

The purpose of the analysis is to find the value of V_L. As a convention, it is assumed that the node voltage is negative with respect to the reference node; the actual polarity is indicated by the sign of the answer. As the result of this convention, the electron current must flow away from the node through the resistors attached to the node. Sources that force current into the node are considered to be positive while those driving current out of the node are negative. The equation will then be: algebraic sum of the source currents entering the node = total of the currents leaving the node through the resistors.

Therefore at the node N,

$$2.55 + 6.0 = \frac{V_L}{4.7} + \frac{V_L}{10} + \frac{V_L}{3.3}$$

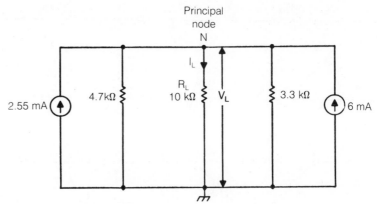

Fig. 8-9 Solution by nodal analysis.

$$V_L \times 0.6158 = 8.55$$

$$V_L = 13.9 \text{ V} \quad \text{and} \quad I_L = \frac{13.9 \text{ V}}{10 \text{ k}\Omega} = 1.39 \text{ mA}$$

Nodal voltage analysis using Kirchhoff's Current Law was able to solve the problem with a single equation while mesh-current analysis based on Kirchhoff's Voltage Law required two simultaneous equations. However, the choice of a particular method depends on the circuit to be analyzed.

DELTA, Y, AND Pi T TRANSFORMATIONS

The arrangement of resistors shown in Fig. 8-10A is known as a delta connection (Δ is the Greek capital letter, delta). The resistors can be rearranged as in Fig. 8-10B so that the formation resembles the Greek letter, π, (Pi). Delta and Pi are therefore alternative names for the same arrangement of components.

Figure 8-11A shows another resistor arrangement that is referred to as a wye (Y) connection. In Fig. 8-11B, the same arrangement has been modified to look like the letter tee (T). Consequently, either wye or tee can be used to signify the same component arrangement.

In many cases, it is impossible to analyze a circuit directly in terms of series, parallel, or series-parallel arrangements. However, if a delta arrangement of resistors can be replaced by an equivalent Y connection (or vice-versa), the resultant circuit sometimes can be changed into a simple series-parallel type.

In order that the delta and wye connections shall be equivalent, it is necessary that the resistances between the points A, B and C in both arrangements shall be the same. For example, in Fig. 8-10A, using the product-over-sum formula, the total resistance between A and B yields

$$\frac{R_{AB}(R_{BC} + R_{CA})}{R_{AB} + R_{BC} + R_{CA}}$$

while in Fig. 8-11A, total resistance between A and B = $R_A + R_B$.

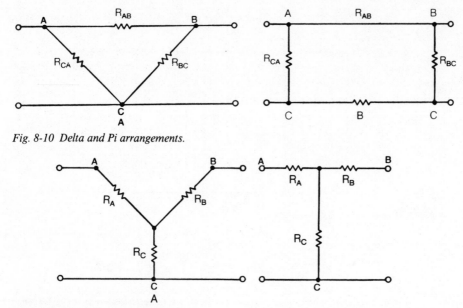

Fig. 8-10 Delta and Pi arrangements.

Fig. 8-11 Wye and tee arrangements.

Therefore:

$$R_A + R_B = \frac{R_{AB}(R_{BC} + R_{CA})}{R_{AB} + R_{BC} + R_{CA}}$$

Two other equations can be obtained for the total resistances between the points B, C and the points C, A.

The results derived from these three equations are

$$R_A = \frac{R_{AB}R_{CA}}{R_{AB} + R_{BC} + R_{CA}}$$

$$R_B = \frac{R_{BC}R_{AB}}{R_{AB} + R_{BC} + R_{CA}}$$

$$R_C = \frac{R_{CA}R_{BC}}{R_{AB} + R_{BC} + R_{CA}}$$

These three equations would be used in transposing from a delta to a wye connection. The next three equations are required for a wye to delta transformation:

$$R_{AB} = \frac{R_A R_B + R_B R_C + R_C R_A}{R_C}$$

$$R_{BC} = \frac{R_A R_B + R_B R_C + R_C R_A}{R_A}$$

$$R_{CA} = \frac{R_A R_B + R_B R_C + R_C R_A}{R_B}$$

A good case for using a delta-wye transformation is in the analysis of a bridge circuit (Fig. 8-12).

As already pointed out, the center 5.6 kΩ resistor is not connected directly in either series or parallel with the other resistors. However, if the points X, Y and Z are regarded as the corners of a delta formation, the equivalent Y connection will result in the circuit of Fig. 8-13. This new circuit is a simple series-parallel arrangement that can be analyzed readily. For the delta-wye transformation using the preceding equations

$$R_X = \frac{5.6 \times 6.8}{5.6 + 6.8 + 3.3} = 2.425 \text{ k}\Omega$$

$$R_Y = \frac{5.6 \times 3.3}{5.6 + 6.8 + 3.3} = 1.177 \text{ k}\Omega$$

$$R_Z = \frac{6.8 \times 3.3}{5.6 + 6.8 + 3.3} = 1.429 \text{ k}\Omega$$

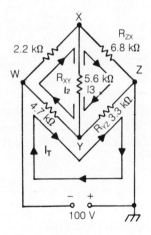

Fig. 8-12 Analysis of bridge circuit.

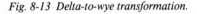

Fig. 8-13 Delta-to-wye transformation.

In Fig. 8-13, the circuit's total resistance is

$$R_T = 1.429 + \frac{(2.2 + 2.425) \times (4.7 + 1.177)}{(2.2 + 2.425) + (4.7 + 1.177)} = 4.017 \text{ k}\Omega$$

$$I_T = \frac{100 \text{ V}}{4.017 \text{ k}\Omega} = 24.89 \text{ mA}$$

The current through the 2.2 kΩ resistor is

$$24.89 \text{ mA} \times \frac{5.877 \text{ k}\Omega}{4.625 \text{ k}\Omega + 5.877 \text{ k}\Omega} = 13.93 \text{ mA}$$

The current through the 4.7 kΩ resistor is

$$24.89 - 13.93 = 10.96 \text{ mA}$$

The potential at point X is $-(100 - 13.93 \times 2.2) = -69.35$ V, while the potential at point Y is $-(100 - 4.7 \times 10.96) = -48.49$ V. The current through the 5.6 kΩ resistor is

$$\frac{69.35 - 48.49}{5.6} = 3.73 \text{ mA}$$

The direction of the electron flow is from point X to point Y.

For comparison purposes, the current through the 5.6 kΩ resistor will be found by other methods.

METHOD 1: Thévenin's Theorem   With this method, the 5.6 Ω resistor is regarded as the load, and therefore the first step is to remove this resistor from the circuit, (Fig. 8-14A). By the voltage division rule, the potential at point X is

$$-100 \text{ V} \times \frac{6.8 \text{ k}\Omega}{2.2 \text{ k}\Omega + 6.8 \text{ k}\Omega} = -75.56 \text{ V}$$

while the potential at point Y is

$$-100 \text{ V} \times \frac{3.3 \text{ k}\Omega}{4.7 \text{ k}\Omega + 3.3 \text{ k}\Omega} = -41.25 \text{ V}$$

Then E_{TH} is $75.56 - 41.25 = 34.31$ V.

The second step is to short out the 100 V source and then obtain the open-circuit resistance between the points X and Y. (Fig. 8-14B). With respect to these points, the 2.2-kΩ and 6.8-kΩ resistors are in parallel; the 4.7-kΩ and 3.3-kΩ resistors also are in parallel, but the two parallel combinations are in series so that R_{TH} is

$$\frac{2.2 \times 6.8}{2.2 + 6.8} + \frac{3.3 \times 4.7}{3.3 + 4.7} = 1.66 + 1.94 = 3.60 \text{ k}\Omega$$

The Thévenin equivalent circuit appears in Fig. 8-14C. When the 5.6 kΩ resistor is replaced, the load current is

$$\frac{34.31 \text{ V}}{3.6 \text{ k}\Omega + 5.6 \text{ k}\Omega} = 3.73 \text{ mA}$$

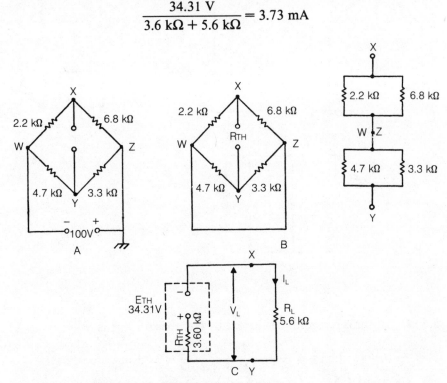

Fig. 8-14 *Analysis of bridge circuit by Thévenin's theorem.*

METHOD 2: Norton's Theorem The first step is to remove the 5.6 kΩ load and replace the resistor by a short circuit (Fig. 8-15A). The total resistance of the circuit is

$$\frac{2.2 \times 4.7}{2.2 + 4.7} + \frac{6.8 \times 3.3}{6.8 + 3.3} = 1.499 + 2.222 = 3.721 \text{ k}\Omega$$

Using the current division rule, the current in the 2.2 kΩ resistor is

$$\frac{100 \text{ V}}{3.721 \text{ k}\Omega} \times \frac{4.7 \text{ k}\Omega}{2.2 \text{ k}\Omega + 4.7 \text{ k}\Omega} = 18.31 \text{ mA}$$

while the current in the 6.8 kΩ resistor is

$$\frac{100 \text{ V}}{3.721 \text{ k}\Omega} \times \frac{3.3 \text{ k}\Omega}{3.3 \text{ k}\Omega + 6.8 \text{ k}\Omega} = 8.78 \text{ mA}$$

Therefore: I_N is $18.31 - 8.78 = 9.53$ mA, and because $R_N = R_{TH}$ (Fig. 8-15B), the equivalent Norton circuit is as shown in Fig. 8-15C. When the

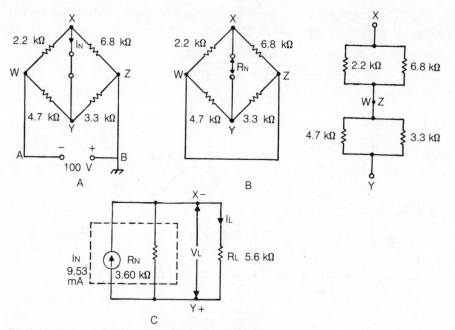

Fig. 8-15 Analysis of bridge circuit by Norton's theorem.

5.6 kΩ resistor is replaced, its current by the division rule is

$$9.53 \text{ mA} \times \frac{3.6 \text{ k}\Omega}{5.6 \text{ k}\Omega + 3.6 \text{ k}\Omega} = 3.73 \text{ mA}$$

METHOD 3. Mesh Current Analysis Referring to Fig. 8-12 the mesh equations are

mesh WYZW: $(I_1 - I_2) 4.7 + (I_1 - I_3) 3.3 = 100$
mesh WXY: $2.2 I_2 + 5.6(I_2 - I_3) + 4.7(I_2 - I_1) = 0$
mesh ZYX: $3.3(I_3 - I_1) + 5.6(I_3 - I_2) + 6.8 I_3 = 0$

Therefore

$$8 I_1 - 4.7 I_2 - 3.3 I_1 = 100$$
$$-4.7 I_1 + 12.5 I_2 - 5.6 I_3 = 0$$
$$-3.3 I_1 - 5.6 I_2 + 15.7 I_3 = 0$$

The solutions to these equations are: $I_1 = 24.89$ mA, $I_2 = 13.93$ mA, and $I = 10.20$ mA. The current through the 5.6 kΩ resistor is: $I_2 - I_3 = 13.93 - 10.20 = 3.73$ mA.

In this case, nodal analysis would have no advantage over mesh analysis because there are three principal nodes at W, X, Y and therefore the number of simultaneous equations is the same for both methods.

CHAPTER SUMMARY

☐ Kirchhoff's Voltage Law

The algebraic sum of the voltages around a closed loop (or mesh) is zero.

☐ Kirchhoff's Current Law

The algebraic sum of the currents at a junction point (or node) is zero

☐ Thévenin's Theorem

Thévenin voltage, E_{TH} = open-circuit voltage between the load terminals (load removed).

Thévenin resistance, R_{TH} = open-circuit resistance between the load terminals (load removed) with all sources replaced by their internal resistances.

☐ Norton's Theorem

Norton current, I_N = short-circuit current between the load terminals (load removed).

Norton resistance, $R_N = R_{TH}$.

☐ Nodal Analysis

At a principal node the algebraic sum of the source currents = the sum of the currents through the resistors.

☐ Delta-to-Wye Transformation

$$R_A = \frac{R_{AB}R_{CA}}{R_{AB} + R_{BC} + R_{CA}}$$

etc.

$$R_{AB} = \frac{R_A R_B + R_B R_C + R_C R_A}{R_C}$$

etc.

☐ Millman's Theorem

Total current of the equivalent constant-current generator = algebraic sum of the currents in the individual constant-current generators.

Internal resistance of the equivalent constant-current generator = total resistance of the constant-current generators' internal resistances in parallel.

9

Introduction to Alternating Current

IN THIS INTRODUCTION to alternating current theory, you will learn about:

☐ The differences between direct and alternating currents.

☐ The use of such terms as peak value, period, frequency, and angular frequency in describing an alternating waveform.

☐ The creation of an alternating sinewave voltage by a simple ac generator.

☐ The meaning of harmonics and their relationship to the fundamental waveform.

☐ The use of the effective or root-mean-square value in measuring the value of an alternating waveform.

☐ The relationship between the effective value, the average value, and the form factor of a sinewave.

☐ The measurement of the phase difference between two alternating waveforms of the same frequency.

☐ The representation of sinusoidal quantities by the use of phasors.

☐ The effect of resistance in a sinewave ac circuit and the phase relationship between the voltage across a resistor and the current flowing through the resistor.

☐ The effect of inductance in a sinewave ac circuit and the phase relationship between the voltage across an inductor and the current flowing through the inductor.

☐ The meaning of inductive reactance as the opposition to alternating current flow provided by an inductor.

☐ The effect of capacitance in a sinewave ac circuit and the phase relationship between the voltage across the capacitor and the current associated with the capacitor.

□ The meaning of capacitive reactance as the opposition to alternating current flow provided by a capacitor.

□ The concept of the reactive power measured in volt-amperes-reactive as compared to the true power, which is measured in watts.

□ The concept of the impedance, which is total opposition to alternating current flow and is a combination of resistance and reactance.

□ The meaning of power factor as the ratio of the true power in watts to the apparent power in volt-amperes.

Previous chapters discussed direct current, in which the electron flow is always in the same direction; however, the magnitude of the flow was not necessarily constant.

COMPARISON BETWEEN DIRECT CURRENT AND ALTERNATING CURRENT

With alternating current, the current reverses direction periodically and has an average value of zero. The current can assume a variety of waveforms, some of which appear in Fig. 9-1. In each case, a single complete waveform is known as a *cycle,* and the number of cycles occuring in one second is the frequency, f. This is the rate at which the waveform repeats and is measured in cycles per second (cs) or hertz (Hz). The time interval taken by a complete cycle is the period, T, which is equal to the reciprocal of the frequency. Therefore

$$t \text{ (seconds)} = \frac{1}{f \text{ (Hz)}}, \text{ and } f = \frac{1}{T} \text{ Hz}$$

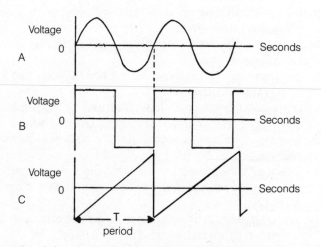

Fig. 9-1 Examples of alternating voltage waveforms; A, sine-wave, B, square-wave, C, saw-tooth-wave.

If the average value of the waveform is not zero, it is regarded as a combination of a dc component and an ac component. The 60-Hz commercial ac line voltage has a sine waveform whose shape is shown in Fig. 9-2. This curve is the result of plotting the mathematical sine function, sin θ, against the angle, θ, measured in either degrees or radians. The radian is the angle at the center of a circle that subtends (is opposite to) an arc equal in length to the radius. Because the total length of the circumference is $2\pi \times$ radius, the angle of 360° must be equivalent to 2π radians. Therefore

$$\theta \text{ (degrees)} = \frac{360}{2\pi} \times \theta \text{ (radians)} = \frac{180}{\pi} \times \theta \text{ (radians)}$$

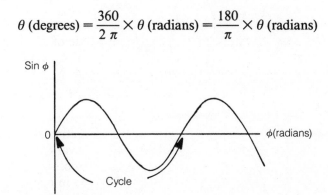

Fig. 9-2 *The mathematical sine-wave.*

and

$$\theta \text{ (radians)} = \frac{\pi}{180} \times \theta \text{ (degrees)}$$

Therefore

$$1 \text{ radian} = \frac{180}{\pi} \text{ degrees} = 57.296° = 57°17'45''$$

Sinewave ac is particularly important because Fourier analysis allows the waveforms in Fig. 9-1 to be broken down in a series of sinewaves that consist of a fundamental component and harmonics (see Fig. 9-3). The fundamental frequency, f, is the same as the waveform's repetition rate. The second harmonic has a frequency 2 f, the third harmonic 3 f, etc, (Fig. 9-4). Compared with other ac waveforms the sinewave also has a high form factor.

An angular velocity of rotation is measured in radians per second and normally is designated by the Greek letter, ω, omega. If the angle of rotation is θ radians, $\theta = \omega t$, where t is the time in seconds.

During a cycle, a total rotation of 2π radians occurs in the time of t seconds. Then the angular frequency is

$$\omega = \frac{\theta}{t} = \frac{2\pi}{T} = 2\pi f \text{ radians per second}$$

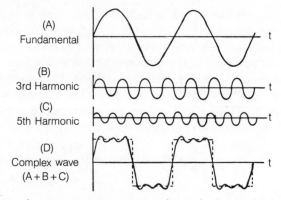

Fig. 9-3 *Synthesis of an approximate square-wave from a fundamental sine-wave and its odd harmonics.*

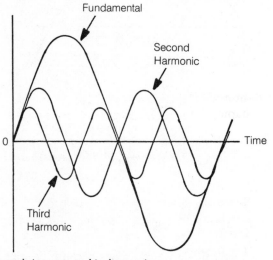

Fig. 9-4 *Fundamental sine-wave and its harmonics.*

THE SIMPLE AC GENERATOR (ALTERNATOR)

The ac generator is a machine that is capable of converting mechanical energy into alternating electrical energy. Figure 9-5A shows a shaft, D, that is driven around by mechanical energy. Attached to the shaft but insulated from it is a loop that rotates between the poles, P, of a permanent magnet or an electromagnet. The ends of the loop are joined to two slip rings, S,S′ which, as they rotate, make contact with stationary carbon brushes, C,C′. These brushes then are connected to the electrical load.

The pole pieces, PP′, are especially shaped to provide a constant flux density in which the loop rotates as shown in Fig. 9-5B.

The loop of n turns moves with an angular velocity of ω radians per second so that the velocity of each of the conductors, A, B, is $v = \omega b$ meters per second, where 2b is the width of the loop in meters. The component of

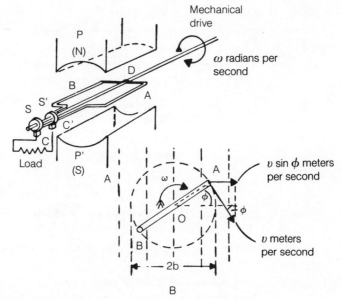

Fig. 9-5 Basic ac generator.

the velocity at right angles to the magnetic flux is $v \sin \phi$. If you use the equations from Chapter 5, the total voltage induced in the loop is

$$e = 2BLnv \sin \phi$$
$$= BAn\omega \sin \omega t \text{ volts}$$

where

B = constant flux density in teslas
L = length of each conductor in meters
$A = 2Lb$ = area of each turn in the loop
t = time in seconds.

Therefore the instantaneous output from the generator is: $e = E_{max} \sin \omega t$, where $E_{max} = BAN\omega$ = the maximum or peak value of the output voltage.

Figure 9-6 shows the waveform of this output voltage. Because the sinewave is symmetrical about the zero line, the peak-to-peak value (E_{p-p}) is equal to twice the peak value of E_{max}. It is an ac voltage's peak-to-peak value that is measured normally on an oscilloscope display.

For the two pole machine of Fig. 9-5A, one complete rotation of the loop generates one cycle of ac through the load. If the output frequency is 60 Hz, the loop's speed of rotation is $60 \times 60 = 3600$ revolutions per minute (rpm); however, if the alternator has four poles, one complete rotation generates two cycles of ac, and therefore the speed required to produce a 60-Hz voltage is only 1800 rpm. The frequency

$$f = \frac{Np}{60} \text{ Hz, and N} = \frac{60 \, f}{p} \text{ rpm}$$

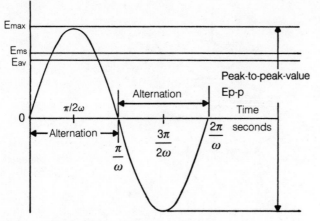

Fig. 9-6 Values associated with the sine-wave.

where

$$p = \text{number of pairs of poles}$$
$$N = \text{speed of rotation in rpm}$$

Example 9-1

A coil of 1000 turns is rotated at 3600 rpm in a magnetic field having a uniform flux of 0.06 tesla. The average area of each turn is 50 cm², and the axis of rotation is at right angles to the direction of the flux. Calculate (a) the frequency, (b) its period, (c) the angular velocity, and (d) the maximum value of the generated voltage. Write down a trigonometric expression for the instantaneous voltage and calculate its value when the coil has rotated 7 radians from the position of zero voltage.

Solution

(a) frequency, $f = \dfrac{3600}{60} = 60$ Hz

(b) period, $t = \dfrac{1}{f} = \dfrac{1}{60}$ second $= 16.67$ milliseconds

(c) angular velocity, $\omega = 2 \pi f = 2 \pi \times 60 = 377$ radians per second

(d) maximum voltage, $E_{max} = BAn\omega$
$$= 0.06 \times 50 \times 10^{-4} \times 10^3 \times 377$$
$$= 113 \text{ V}$$

Therefore the instantaneous voltage,

$$e = E_{max} \sin \omega t$$
$$= 113 \sin 377\ t$$
$$= 113 \sin \phi$$

If

$$\phi = 7 \text{ radians,}$$
$$e = 113 \sin (7 \text{ radians})$$
$$= 74 \text{ V}$$

EFFECTIVE OR ROOT-MEAN-SQUARE (rms) VALUE

In Fig. 9-7A, an ac voltage of instantaneous voltage $e = E_{max} \sin \omega t$ is applied across resistor, R. The instantaneous power dissipated is

$$p = \frac{e^2}{R} = \frac{E_{max}^2 \sin^2 \omega t}{R} = \frac{E_{max}^2 (1 - \cos 2 \omega t)}{2 R}$$

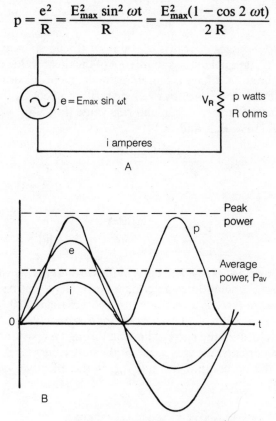

A

B

Voltage, current, and power waveforms

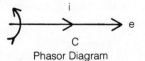

C

Phasor Diagram

Fig. 9-7 Resistance in ac circuits.

The instantaneous power ranges from zero to a peak value of E_{max}^2/R. The average value of cos 2 ωt over a complete cycle is zero; this is indicated by the symmetry of the instantaneous power curve in Fig. 9-7B.

The average power is therefore

$$P_{av} = \frac{E_{max}^2}{2\,R} = \frac{\left(\dfrac{E_{max}}{\sqrt{2}}\right)^2}{R}$$

The quantity

$$\frac{E_{max}}{\sqrt{2}} = \frac{E_{max}}{1.414} = 0.707\ E_{max}$$

is called the *root-mean-square* (rms) value of the ac sinewave voltage. In other words, E_{rms} is equal to the square root of the mean value of e^2. The rms value of an alternating voltage is therefore the steady or constant voltage that would provide the same mean power or heating effect as the alternating voltage. To convert between peak and rms value (Fig. 9-6), E_{peak} or $E_{max} = \sqrt{2} \times E_{rms} = 1.414 \times E_{rms}$ and

$$E_{rms} = \frac{E_{peak}}{\sqrt{2}} = \frac{E_{peak}}{1.414} = 0.707 \times E_{peak}$$

Similarly,

$$I_{rms} = \frac{I_{peak}}{\sqrt{2}}$$

etc.

Most alternating voltage and current measurements are in terms of rms values. However, insulation in ac circuits normally must be able to withstand the peak voltage. For example, the commercial 110 V, 60-Hz line voltage has a peak value of $110 \times \sqrt{2}$, or approximately 155 V.

If the waveform of the current flowing through a resistor contains both dc and ac components, the rms value of the entire current is $\sqrt{I_{DC}^2 + I_{rms}^2}$, where I_{DC} is the dc component and I_{rms} is the effective value of the ac component (Fig. 9-8).

AVERAGE VALUE AND FORM FACTOR

The full cycle of a sinewave is composed of two alternations: one alternation extends from 0° to 180°, and the other from 180° to 360° (Fig. 9-6). Although the average value of the sinewave over the complete cycle is zero, the average value over an alternation is

$$E_{av} = \frac{\displaystyle\int_o^{\pi/\omega} E_{max} \sin \omega t,\ dt}{\displaystyle\int_o^{\pi/\omega} dt}$$

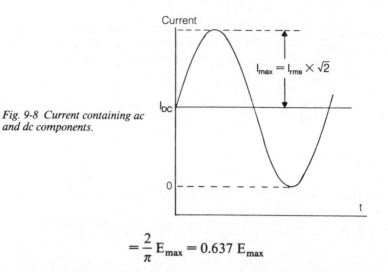

Current

$I_{max} = I_{rms} \times \sqrt{2}$

I_{DC}

0

t

Fig. 9-8 Current containing ac
and dc components.

$$= \frac{2}{\pi} E_{max} = 0.637\ E_{max}$$

The average value over an alternation appears in Fig. 9-6.

The ratio of the rms value to the average value of an alternating current or voltage is called the *"form factor"* of the waveform. For the sinewave,

$$\text{form factor} = \frac{0.707\ E_{max}}{0.637\ E_{max}} = 1.11$$

Compared with other ac waveforms, the sinewave is relatively easy to generate and has a superior form factor.

PHASE

If two dc voltages are in series, either they are aiding or opposing each other so that you only need to add or subtract in order to obtain the total voltage. However, two ac waveforms of the same frequency may not reach similar points in the cycle, for example, peak or zero values, at the same time. The amount by which the two waveforms are out of step is referred to as their *phase difference,* which is measured in either degrees or radians.

In Fig. 9-10A the e_2 waveform reaches its peak, X, earlier in time than the e waveform with its corresponding peak, Y. The phase difference between the two waveforms is ϕ radians with e_2 leading e (or e_1 lagging e_2). The waveform equations are

$$e_2 = E_{2max} \sin(\omega t + \phi)$$

and

$$e_1 = E_{1max} \sin \omega t$$

Similarly, the equation $e_3 = E_{3max} \sin(\omega t - \phi)$ means that e_3 lags e_1 (or e_1 leads e_3) by ϕ radians; also e_2 leads e_3 by 2ϕ radians.

Phase differences therefore can either lead or lag, and the angles usually extend up to 180°. In the particular case of $\phi = 180°$ or π radians, the terms

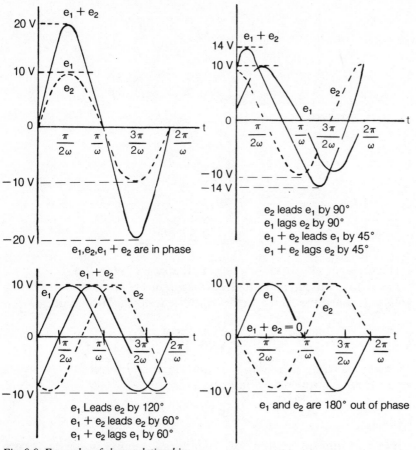

Fig. 9-9 *Examples of phase relationships.*

"leading" or "lagging" are not used because 180° lagging has the same meaning as 180° leading. In most cases, there is little point in using angles greater than 180°, because, for example, 270° lagging is equivalent to 90° leading.

It is impossible to add together two or more ac voltages (or currents if some form of parallel circuit is involved) without knowing their magnitudes and their phase relationship. For example, if a resistor and an inductor are connected in a series ac circuit, their voltages are 90°, or $\pi/2$ radians, out of phase, and you must take this phase relationship into account when adding the voltages. Figure 9-9 shows the result of adding two ac voltages, each of 10 volts peak, but with phase differences that are 0°, 90°, 120°, and 180°. The corresponding resultant voltages have peak values of 20 V, 14.14 V, 10 V, and zero.

In the general case of $e_1 = E_{1max} \sin \omega t$ and $e_2 = E_{2max} \sin (\omega t + \phi)$, the sum of the two ac voltages is

$$e_1 + e_2 = E_{1max} \sin \omega t + E_{2max} \sin (\omega t + \phi)$$

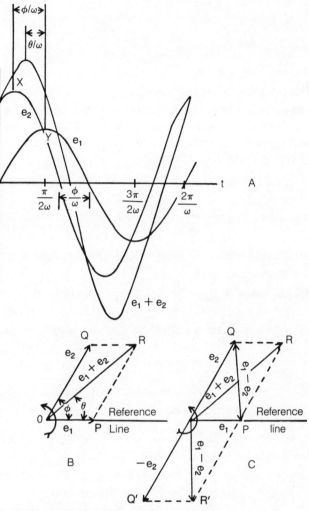

Fig. 9-10 Addition and subtraction of the two sine-wave voltages.

$$= E_{1max} \sin \omega t + E_{2max} \cos \phi \sin \omega t + E_{2max} \sin \phi \cos \omega t$$
$$= \sqrt{E_{1max}^2 + E_{2max}^2 + 2\,E_{1max}E_{2max} \cos \phi}\, \sin (\omega t + \theta)$$

where:

$$\theta = \text{inv. (inverse) tan. or arc. tan. or } \tan^{-1}\left(\frac{E_{2max} \sin \phi}{E_{1max} + E_{2max} \cos \phi}\right).$$

The resultant voltage, $e_1 + e_2$, therefore has the same frequency as e_1 and e_2. It has a peak value of $\sqrt{E_{1max}^2 + E_{2max}^2 + 2\,E_{1max}E_{2max} \cos \phi}$ and leads e_1 by ϕ radians (Fig. 9-10A).

It follows that:

$$e_1 - e_2 = \sqrt{E_{1max}^2 + E_{2max}^2 - 2\,E_{1max}E_{2max} \cos \phi}\, \sin (\omega t + \theta)$$

where

$$\theta = \text{inv. tan.} \left(\frac{E_{2max} \sin \phi}{E_{2max} \cos \phi - E_{1max}} \right)$$

Example 9-2

A sinewave voltage of peak-to-peak value 24 V is applied across an 80 Ω resistor. Calculate (a) the peak, average (over an alternation), and effective values of the sinewave voltage, and (b) the values of the peak power and the average power dissipated over the cycle.

Solution

(a) peak value, $E_{max} = \dfrac{E_{p-p}}{2} = \dfrac{24}{2} = 12$ V

average value, $E_{av} = \dfrac{2}{\pi} \times 12 = 0.637 \times 12 = 7.64$ V

effective value, $E_{rms} = \dfrac{12}{\sqrt{2}} = 0.707 \times 12 = 8.49$ V

(b) peak power $= \dfrac{E_{max}^2}{R} = \dfrac{(12 \text{ V})^2}{80 \ \Omega} = \dfrac{144}{80} = 1.8$ W

average power $= \dfrac{\text{peak power}}{2} = \dfrac{1.8}{2} = 0.9$ W

alternatively, average power $= \dfrac{E_{rms}^2}{R} = \dfrac{8.49^2}{80} = 0.9$ W

Example 9-3

A sinewave voltage has an effective value of 110 V. Calculate its peak to-peak value and its average value over an alternation.

Solution

$$\text{peak-to-peak value} = \sqrt{2} \ E_{rms} = 2.828 \times E_{rms}$$
$$= 2.828 \times 110$$
$$= 311 \text{ V}$$

$$\text{average value over an alternation} = \frac{2 \sqrt{2}}{\pi} \times E_{rms}$$
$$= 0.9 \times 110 = 99 \text{ V}$$

Example 9-4

Two alternating voltages are represented by $e_1 = 10 \sin \omega t$ and $e_2 = 15 \sin (\omega t + \pi/6)$. What is the phase relationship between e_1 and e_2? What are the trigonometrical expressions for $e_1 + e_2$ and $e_1 - e_2$? Verify these results by drawing to scale the sinewaves for e_1, e_2, $e_1 + e_2$, $e_1 - e_2$ and $e - e_1$.

Solution

Because $\pi/6$ radians $= 30°$, e_1 lags e_2 by $30°$ or e_2 leads e_1 by $30°$

$$e_1 + e_2 = \sqrt{10^2 + 15^2 + 2 \times 10 \times 15 \times \cos 30°} \, \sin(\omega t + \phi_1)$$

where

$$\phi_1 = \text{inv. tan.} \left(\frac{15 \sin 30°}{10 + 15 \cos 30°} \right) = 18.1°$$

Therefore $e_1 + e_2 = 24.18 \sin(\omega t + 18.1°)$, which leads e_1 by $18.1°$ but lags e_2 by $30° - 18.1° = 11.9°$.

$$e_1 - e_2 = \sqrt{10^2 + 15^2 - 2 \times 10 \times 15 \times \cos 30°} \, \sin(\omega t + \phi_2)$$

where

$$\phi_2 = \text{inv. tan.} \left(\frac{15 \sin 30°}{15 \cos 30° - 10} \right) = -111.7°$$

Therefore $e_1 - e_2 = 8.07 \sin(\omega t - 111.7°)$, which leads e_1 by $68.3°$ and e_2 by $38.3°$.

Note that $e_2 - e_1 = 8.07 \sin(\omega t + 68.3°)$, which is $180°$ out of phase with $e_1 - e_2$.

The waveforms for e_1, e_2, $e_1 + e_2$ and $e_2 - e_1$, appear in Fig. 9-11.

REPRESENTATION OF AN AC VOLTAGE OR CURRENT BY A ROTATING PHASOR

So far in this chapter an ac voltage or current has been represented either by a trigonometrical equation or graphically by a waveform. Both these methods tend to be tedious, particularly if voltages (or currents) have to be added or subtracted.

A third method involves phasor representation. If the equation of an ac voltage is $e = E_{max} \sin \omega t$, its phasor is a line whose length is equal to the peak value, E_{max}. By convention, the phasor is assumed to rotate in the positive or counter-clockwise direction with an angular velocity of ω radians per second. The vertical projection of the phasor on the horizontal reference line equals the instantaneous value of the ac voltage (Fig. 9-12). Therefore as the phasor rotates, its extremity can be said to trace out the ac voltage's sinusoidal waveform with a frequency equal to the phasor's speed of rotation in revolutions per second. The phasor diagram therefore contains the same information as the waveform presentation, and it is obviously easier to work with lines rather than sinewaves.

Because there is a constant relationship between rms and peak values, (rms value $= 0.707 \times$ peak value), the length of the phasor can be used to indicate the rms value.

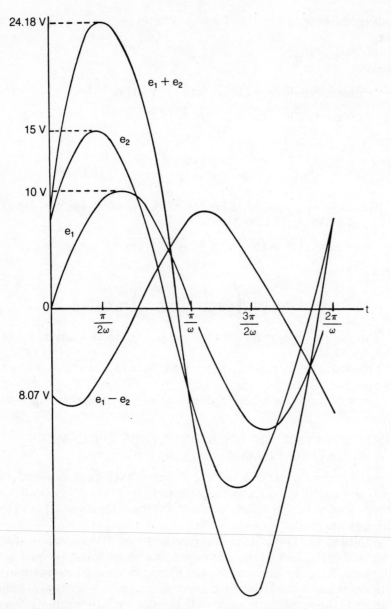

Fig. 9-11 Waveforms related to example 9-4.

If two sinewaves are not in phase, the phase difference are represented by the angular separation between their phasors. In Fig. 9-10B, the phasors for $e_1 = E_{1max} \sin \omega t$ and $e_2 = E_{2max} \sin (\omega t + \phi)$ are separated by an angle of ϕ radians, with e_2 leading e_1 as the phasors rotate in the counter-clockwise direction. In many phasor diagrams one of the phasors, in this case e_1, is chosen as a reference and is placed along the horizontal line.

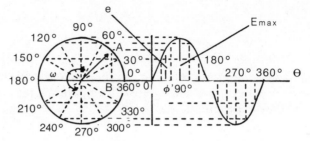

Fig. 9-12 Phasor representation of a sine-wave.

ADDITION AND SUBTRACTION OF PHASORS

Previous sections have shown how to add together two sinewaves either trigonometrically or by a waveform presentation. In order to add the phasors e_1 and e_2, it is necessary to construct the parallelogram of Fig. 9-10B and then draw the diagonal OR to represent the phasor, $e_1 + e_2$. It can then be shown that the length, OR, is $\sqrt{E_{1max}^2 + E_{2max}^2 + 2\,E_{1max}E_{2max}\cos\phi}$ and that the angle, ϕ, is inv. (arc) tan $E_{2max} \sin\phi/(E_{1max} + E_{2max}\cos\phi)$. These are exactly the same results that were obtained by adding the sinewaves trigonometrically and this therefore justifies the use of the parallelogram of phasors for the operation of addition.

To obtain $e_1 - e_2 = e_1 + (-e_2)$ in the process of subtracting phasors, method involves finding $(-e_2)$, which is then added to e_1. Because $-e_2 = -E_{2max} \sin(\omega t + \phi) = E_{2max} \sin(\omega t + \phi + \pi)$, the negative sign causes the phasor to be rotated through π radians or 180° as shown in Fig. 9-10C. Adding the phasors e_1 and $-e_2$ together produces the diagonal OR′, which represents $e_1 - e_2$ and is also the same as the other diagonal of the parallelogram OPQR. In other words, one of the diagonals, OR, represents the phasor sum, and the other diagonal, QP, is the phasor difference.

It is worth mentioning that prior to about 1960 it was common practice to refer to phasors as vectors. A vector is a quantity that possesses both magnitude and direction while a scalar possesses a magnitude only. Therefore force and velocity are vectors while power is a scalar. The vectors in mechanics obey the same rules as phasors for addition and subtraction but the rules for multiplication and division are totally different; to emphasize this distinction, the term "phasor" was introduced. To establish the rules for multiplication and division of phasors will require the use of complex algebra, which is discussed in chapter 13.

Now that we have the phasor, waveform, and trigonometrical representations of sinewave voltages and currents, we can examine the effects of resistance, inductance, and capacitance in ac circuits.

Example 9-5

Two alternating voltages are represented by

$$e_1 = 10 \sin \omega t$$

and

$$e_2 = 15 \sin\left(\omega t + \frac{\pi}{6}\right)$$

Draw to scale a phasor diagram containing e_1, e_2, and construct the phasors representing $e_1 + e_2$, $e_1 - e_2$, and $e_2 - e_1$. Compare the results with the answers to the previous example.

Solution

The relevant phasor diagram is shown in Fig. 9-13.
Within the limits of measurement, the phasors are

$$e_1 + e_2 = 24.2 \sin(\omega t + 18°)$$
$$e_1 - e_2 = 8.1 \sin(\omega t - 112°)$$
$$e_2 - e_1 = 8.1 \sin(\omega t + 68°)$$

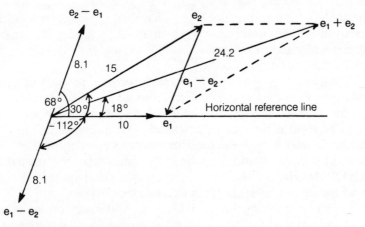

Fig. 9-13 Phasor diagram for example 9-5.

RESISTANCE IN AN AC CIRCUIT

Let an instantaneous ac voltage, $e = E_{max} \sin \omega t$, be applied to a resistor, R (Fig. 9-7); note that $v_R = -E_{max} \sin \omega t$ and $e + v_R = 0$. Because Ohm's Law applies at any instant, the instantaneous current is

$$i = \frac{e}{R} = \frac{E_{max} \sin \omega t}{R} = I_{max} \sin \omega t,$$

e and i are therefore in phase as shown in the waveform and phasor diagrams. Also,

$$\frac{E_{max}}{I_{max}} = \frac{1.414\, E_{rms}}{1.414\, I_{rms}} = \frac{E_{rms}}{I_{rms}} = R$$

The instantaneous power, p, is

$$p = ei = E_{max}I_{max} \sin^2 \omega t$$
$$= \frac{E_{max}I_{max}}{2} (1 - \cos 2 \omega t)$$

The average power over the cycle is

$$\frac{E_{max}I_{max}}{2} = \frac{E_{max}}{\sqrt{2}} \times \frac{I_{max}}{\sqrt{2}} = E_{rms} \times I_{rms}$$

The ac formulas for resistances in series and parallel are the same as those for dc.

Example 9-6

In Fig. 9-7, e = 18 sin ($\omega t + \pi/4$) volts, and R = 6.8 kΩ. Calculate the effective values of the current, the peak power, and the average power over the cycle. Write down the trigonometrical expression for the instantaneous current.

Solution

$$\text{effective voltage, } E_{rms} = \frac{E_{max}}{\sqrt{2}} = 18 \times 0.707 = 12.73 \text{ V}$$

$$\text{effective current, } I_{rms} = \frac{E_{rms}}{R} = \frac{12.73 \text{ V}}{6.8 \text{ k}\Omega} = 1.87 \text{ mA}$$

$$\text{peak power} = \frac{E_{max}^2}{R} = 47.6 \text{ mW}$$

$$\text{average power} = E_{rms} \times I_{rms} = 23.8 \text{ mW}$$

Because e and i are in phase, the instantaneous current, is

$$i = \frac{18}{6.8} \sin \left(\omega t + \frac{\pi}{4} \right)$$
$$= 2.65 \sin \left(\omega t + \frac{\pi}{4} \right) \text{ mA}$$

INDUCTANCE IN AN AC CIRCUIT

Let an instantaneous ac voltage, e = $E_{max} \sin \omega t$, be applied to an inductor L (Fig. 9-14). The inductor is regarded as possessing inductance only and therefore its resistance is zero.

The induced voltage or counter EMF, $v_L = -L \times$ rate of change of current $= -L \, di/dt$, using calculus notation.

By Kirchhoff's Voltage Law

$$e + v_L = 0$$

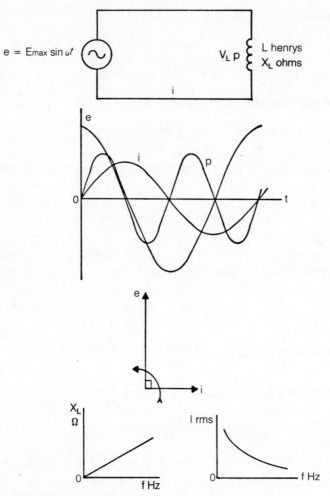

Fig. 9-14 Inductance in ac circuits.

and therefore

$$e = L \frac{di}{dt}$$

so that

$$\frac{di}{dt} = \frac{e}{L} = \frac{E_{max} \sin \omega t}{L}.$$

Integrating with respect to t,

$$i = -\frac{E_{max} \cos \omega t}{\omega L} = \frac{E_{max} \sin \left(\omega t - \frac{\pi}{2} \right)}{\omega L}$$

$$= I_{max} \sin \left(\omega t - \frac{\pi}{2} \right).$$

i therefore lags e by $\pi/2$ radians, or $90°$. Alternatively e leads i by $90°$, and this can be remembered by the word "eLi" (for an inductor L, the instantaneous voltage, e, leads the instantaneous current, i, by $90°$). When voltage leads current, the phase angle is considered to be positive and therefore $\phi = +90°$. Also,

$$\frac{E_{max}}{I_{max}} = \frac{E_{rms}}{I_{rms}} = \omega L = 2 \pi fL = X_L \ \Omega.$$

Therefore

$$I_{rms} = \frac{E_{rms}}{X_L} = \frac{E_{rms}}{2 \pi fL}, \ f = \frac{X_L}{2 \pi L} \ Hz \ \text{and} \ L = \frac{X_L}{2 \pi f} \ H$$

The quantity X_L is called the *inductive reactance* and measures the inductor's total opposition to alternating current.

Note that X_L is directly proportional to frequency while I_{rms} is inversely proportional to frequency. These relationships are illustrated in Fig. 9-14.

The instantaneous power in the circuit is

$$p = ei = E_{max} \sin \omega t \times I_{max} \sin \left(\omega t - \frac{\pi}{2} \right)$$

$$= \frac{E_{max} I_{max}}{2} \sin 2 \left(\omega t - \frac{\pi}{2} \right)$$

$$= E_{rms} \times I_{rms} \sin \left[2 \left(\omega t - \frac{\pi}{2} \right) \right]$$

The mean value of $\sin [2 (\omega t - \pi/2)]$ is zero, and therefore the average value of the power over the complete cycle is also zero (Fig. 9-14). Once again the frequency of the instantaneous power curve is a second harmonic. During two quarters of the cycle, energy is taken from the source in establishing a magnetic flux around the inductor while during the remainder of the cycle, the magnetic flux collapses and energy is returned to the source. This illustrates the difference between resistance and reactance. Both resistance and reactance limit the flow of alternating current, but while resistance dissipates power, reactance does not.

The product $E_{rms} I_{rms}$ is called the idle, wattless or reactive power and is measured in volt-amperes-reactive (VArs).

Now consider inductive reactances in series. The equation for inductors in series is: $L_T = L_1 + L_2 + L_3 + \cdots\cdots + L_N$. Therefore: $2 \pi fL_T = 2 \pi fL_1 + 2 \pi fL_2 + 2 \pi fL_3 + \cdots\cdots + 2 \pi fL_N$ or $X_{LT} = X_{L1} + X_{L2} + X_{L3} + \cdots\cdots + X_{LN}$.

The total inductive reactance is equal to the sum of the individual reactances.

Similarly, consider inductive reactances in parallel. The following equation is for inductors in parallel:

$$\frac{1}{L_T} = \frac{1}{L_1} + \frac{1}{L_2} + \frac{1}{L_3} + \cdots + \frac{1}{L_N}$$

Therefore

$$\frac{1}{2\pi f L_T} = \frac{1}{2\pi f L_1} + \frac{1}{2\pi f L_2} + \frac{1}{2\pi f L_3} + \cdots + \frac{1}{2\pi f L_N}$$

$$\frac{1}{X_{LT}} = \frac{1}{X_{L1}} + \frac{1}{X_{L2}} + \frac{1}{X_{L3}} + \cdots + \frac{1}{X_{LN}}$$

The formulas for inductive reactances and resistances are therefore the same.

Example 9-7

In Fig. 9-14, $e = 12 \sin(5000\,t + \pi/3)$ volts and $L = 850$ mH. Calculate the values of the effective current and the reactive power. Write down the trigonometrical expression for the instantaneous current. If the frequency is doubled, what is the new value of the effective current?

Solution

$$\text{effective applied voltage, } E_{rms} = \frac{12}{\sqrt{2}} = 8.49 \text{ V}$$

angular velocity, $\omega = 5000$ radians per second

inductive reactance, $X_L = \omega L = 5000 \times 850 \times 10^{-3} = 3250\ \Omega$

$$\text{effective current, } I_{rms} = \frac{E_{rms}}{X_L} = \frac{8.49 \text{ V}}{3250\ \Omega} = 2.61 \text{ mA}$$

average reactive power $= E_{rms} \times I_{rms} = 22.16$ mVArs

Because e leads i by 90° or $\pi/2$ radians, the instantaneous current is

$$i = 2.61 \times 1.414 \sin\left(5000\,t + \frac{\pi}{3} - \frac{\pi}{2}\right) = 3.69 \sin\left(5000 - \frac{\pi}{6}\right) \text{ mA}$$

If the frequency is doubled, the inductive reactance is doubled and the effective current is halved. The new effective current is therefore

$$\frac{2.61}{2} = 1.305 \text{ mA}$$

Example 9-8

A 750-kHz sinewave voltage whose effective value is 5 V, is applied across an inductor. If the effective current is 1.75 mA, calculate the value of the inductor.

Solution

$$\text{inductive reactance, } X_L = \frac{E_{rms}}{I_{rms}} = \frac{5 \text{ V}}{1.75 \text{ mA}} = 2857 \ \Omega$$

$$\text{inductance, } L = \frac{X_L}{2 \ \pi f} = \frac{2857}{2 \times \pi \times 750 \times 10^3} \text{ H} = 0.606 \text{ mH}$$

Example 9-9

Three inductors whose reactances are 150 Ω, 275 Ω, and 310 Ω, are connected in series across a 110 V (rms), 60-Hz source. Calculate the effective current and the individual voltages across the inductors. What is the value of the total inductance?

Solution

$$\text{total inductive reactance, } X_{LT} = 150 + 275 + 310 = 735 \ \Omega$$

$$\text{effective current} = \frac{110 \text{ V}}{735 \ \Omega} = 0.1496 \text{ A}$$

The individual voltages across the inductors are 150 Ω × 0.1496 A = 22.45 V, 275 Ω × 0.1496 A = 41.16 V and 310 Ω × 0.1496 A = 46.39 V.

$$\text{total inductance} = \frac{X_{LT}}{2 \ \pi f} = \frac{735}{2 \times \pi \times 60} = 1.95 \text{ H}$$

CAPACITANCE IN AN AC CIRCUIT

Let an ac voltage $e = E_{max} \sin \omega t$ be applied to a capacitor C (Fig. 9-15). Then $v_C = -q/C$ and $e + v_C = 0$. Therefore

$$e = q/C \quad \text{and} \quad q = Ce = CE_{max} \sin \omega t$$

Differentiating with respect to t,

$$i = \frac{dq}{dt} = \omega CE_{max} \cos \omega t$$

$$= \omega CE_{max} \sin \left(\omega t + \frac{\pi}{2} \right) = I_{max} \sin \left(\omega t + \frac{\pi}{2} \right)$$

e therefore lags i by $\pi/2$ radians, or 90° so that the phase angle, ϕ, is −90°. Alternatively, i leads e by 90°, which can be remembered by the word "iCe"; this may be combined with "eLi" to form "eLi, the iCe man".

$$\text{Because } \omega CE_{max} = I_{max}, \frac{E_{max}}{I_{max}} = \frac{E_{rms}}{I_{rms}} = \frac{1}{\omega C} = \frac{1}{2 \ \pi f C} = X_C \ \Omega$$

Then:

$$I_{rms} = 2 \ \pi f C E_{rms}$$

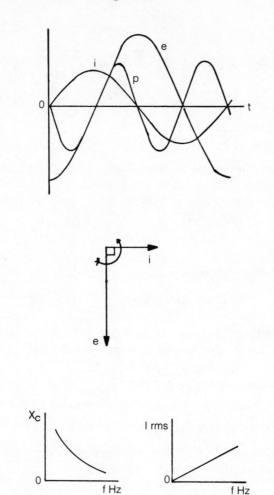

Fig. 9-15 Capacitance in ac circuits.

and

$$C = \frac{1}{2\pi f X_C} \; F$$

$$f = \frac{1}{2\pi C X_C} \; Hz.$$

The quantity X_C is called the *capacitive reactance,* measures the opposition to alternating current offered by a capacitor. Notice that X_C is inversely proportional to f so that the graphs of I_{rms} and X_C against frequency are as shown in Fig. 9-15. Because X_L is directly proportional to frequency while X_C is inversely proportional to frequency, there must be, for particular values of L and C, a certain frequency for which $X_L = X_C$.

The instantaneous power is

$$p = ei = E_{max}I_{max} \sin \omega t \sin \left(\omega t + \frac{\pi}{2} \right)$$

$$= \frac{E_{max}I_{max}}{2} \sin 2 \omega t$$

The average power over the complete cycle is zero. During two quarters of the cycle, energy is taken from the source in charging the capacitor, but during the remainder of the cycle, the capacitor discharges and the energy is returned to the source.

Now consider capacitive reactances in series. The following equation is for N capacitors in series:

$$\frac{1}{C_T} = \frac{1}{C_1} + \frac{1}{C_2} + \frac{1}{C_3} + \cdots + \frac{1}{C_N}$$

Then

$$\frac{1}{2 \pi f C_T} = \frac{1}{2 \pi f C_1} + \frac{1}{2 \pi f C_2} + \frac{1}{2 \pi f C_3} + \cdots + \frac{1}{2 \pi f C_N}$$

or

$$X_{CT} = X_{C1} + X_{C2} + X_{C3} + \cdots X_{CN}$$

Although the reciprocal formula is required for capacitances in series, the total capacitive reactance is merely the sum of the individual reactances.

Now consider capacitive reactances in parallel. The following equation is for N capacitors in parallel

$$C_T = C_1 + C_2 + C_3 + \cdots + C_N$$

Then

$$\frac{2 \pi f C_T}{1} = \frac{2 \pi f C_1}{1} + \frac{2 \pi f C_2}{1} + \frac{2 \pi f C_3}{1} + \cdots + \frac{2 \pi f C_N}{1}$$

or

$$\frac{1}{X_{CT}} = \frac{1}{X_{C1}} + \frac{1}{X_{C2}} + \frac{1}{X_{C3}} + \cdots + \frac{1}{X_{CN}}$$

Although capacitances in parallel are simply added, the reciprocal or repeated product-over-sum formula is required to find the total capacitive reactance.

Example 9-10

In Fig. 9-15, $e = 17 \sin (2 \pi \times 3.25 \times 10^3 t - \pi/12)$ volts and $C = 0.15 \ \mu F$. Calculate the value of the rms current and write down the trigonometrical expression for the instantaneous current. If a further 0.33-μf capacitor is added in parallel with C and the frequency is then tripled, what is the new value of the effective current?

Solution

Frequency, $f = 3.25 \times 10^3$ Hz, and the effective applied voltage is $17/\sqrt{2} = 12.02$ V.

$$\text{capacitive reactance, } X_C = \frac{1}{2\pi fC}$$

$$= \frac{1}{2 \times \pi \times 3.25 \times 10^3 \times 0.15 \times 10^{-6}}$$

$$= 326 \ \Omega$$

$$\text{effective current, } I_{rms} = \frac{E_{rms}}{X_C} = \frac{12.02 \text{ V}}{326 \ \Omega} = 36.83 \text{ mA}$$

and i leads e by $\pi/2$ radians; therefore the instantaneous current is

$$i = \frac{17}{326} \sin \left(2\pi \times 3.25 \times 10^3 \, t - \frac{\pi}{12} + \frac{\pi}{2} \right) \text{ A}$$

$$= 52.1 \sin \left(2\pi \times 3.25 \times 10^3 \, t - \frac{5\pi}{12} \right) \text{ mA}$$

When the 0.33-μF capacitor is added in parallel, the total capacitance, C_T, is $0.33 + 0.15 = 0.48 \ \mu$F. The frequency is changed to $3 \times 3.25 \times 10^3 = 9.75 \times 10^3$ Hz, so that

$$X_{CT} = \frac{1}{2\pi \times 9.75 \times 10^3 \times 0.48 \times 10^{-6}} = 34 \ \Omega$$

The new effective current is $12.02/34 = 0.354$ A.

Example 9-11

Three capacitors with values of 150 pF, 220 pF, and 180 pF are connected in series across a 12 V rms, 1 MHz source. Calculate the individual voltages across the capacitors.

Solution

The individual capacitive reactances are

$$X_{C1} = \frac{1}{2\pi fC_1} = \frac{1}{2 \times \pi \times 1 \times 10^6 \times 150 \times 10^{-12}} = 1061 \ \Omega$$

$$X_{C2} = X_{C1} \times \frac{C_1}{C_2} = 1061 \times \frac{150}{220} = 723 \ \Omega$$

$$X_{C3} = 1061 \times \frac{150}{180} = 884 \ \Omega$$

total capacitive reactance, $X_{CT} = 1061 + 723 + 884 = 2668 \ \Omega$. Alternatively,

$$\frac{1}{C_T} = \frac{1}{150} + \frac{1}{220} + \frac{1}{180}$$
$$= 0.00667 + 0.00455 + 0.00556$$
$$= 0.016767$$

and

$$X_{CT} = \frac{1}{2 \ \pi f C_T} = \frac{0.016767 \times 10^{12}}{2 \times \pi \times 1 \times 10^6} = 2668 \ \Omega$$

Using the voltage division rule,

$$V_{C1} = 12 \ V \times \frac{1061 \ \Omega}{2668 \ \Omega} = 4.77 \ V$$

$$V_{C2} = 12 \ V \times \frac{723 \ \Omega}{2668 \ \Omega} = 3.25 \ V$$

$$V_{C3} = 12 \ V \times \frac{884 \ \Omega}{2668 \ \Omega} = 3.98 \ V$$

Check:

$$E = V_{C1} + V_{C2} + V_{C3} = 12.00 \ V$$

THE GENERAL AC CIRCUIT

The general ac circuit will contain resistance, inductive reactance, and capacitive reactance. The total opposition to the flow of alternating current is therefore a resistance/reactance combination that is called the *impedance*, z.

The impedance, z is a phasor which is defined as the ratio of the source voltage phasor to the source current phasor. The magnitude of z is measured in ohms and is given by

$$Z = \frac{E_{rms}}{I_{rms}} = \frac{E_{max}}{I_{max}}$$

where E_{rms} and I_{rms} are the effective values of the source voltage and the source current.

The direction of z is measured by the angle, ϕ, which is between the z phasor and the horizontal reference line. The value of this angle normally extends from $-90°$ (capacitive reactance only) through $0°$ (resistance only) to $+90°$ (inductive reactance only).

Because the source voltage and the source current are $\phi°$ out of phase, the current can be represented by $i = I_{max} \sin \omega t$, and the voltage will then be

$e = E_{max} \sin (\omega t \pm \phi)$. The instantaneous power is

$$p = ei = E_{max}I_{max} \sin \omega t \sin (\omega t \pm \phi).$$

$$= \frac{E_{max}I_{max}}{2} (\cos \phi - \cos (2 \omega t \pm \phi))$$

The mean value of $\cos (2 \omega t \pm \phi)$ is zero, and therefore the average power over the complete cycle is

$$\frac{E_{max}I_{max} \cos \phi}{2} = E_{rms} \times I_{rms} \cos \phi \text{ watts}$$

This is the true power dissipated in the resistive part of the impedance. By contrast, the product $E_{rms} \times I_{rms}$ is called the *apparent* power at the source and is measured in voltamperes (VA). The cosine of the phase angle, $\cos \phi$, is called the *power factor* and is defined by:

$$\text{power factor, } \cos \phi = \frac{\text{true power, watts}}{\text{apparent power, volt-amperes}}$$

or

$$\text{true power} = E_{rms} \times I_{rms} \times \text{power factor}$$

If an ac current is inductive, the phase angle is positive, and the power factor is lagging (source current lags source voltage). In a capacitive circuit, the phase angle is negative and the power factor is leading. The value of the power factor extends from zero (reactance only) to 1 (resistance only).

Example 9-12

An alternating source has an instantaneous voltage of $e = 28 \sin (\omega t + \pi/4)$ volts and is connected across an ac circuit. If the instantaneous current drawn from the source is

$$5.7 \sin \left(\omega t - \frac{\pi}{12} \right) \text{mA}$$

calculate the values of the impedance, the phase angle, and the power factor. What are the values of the true power and the apparent power?

Solution

$$\text{impedance, } Z = \frac{E_{rms}}{I_{rms}} = \frac{28 \text{ V}}{5.7 \text{ mA}} = 4.91 \text{ k}\Omega$$

e leads i by

$$\phi = \frac{\pi}{4} - \left(-\frac{\pi}{12} \right) = +\frac{\pi}{3} \text{ radians}$$

the power factor is inv $\cos\left(+\dfrac{\pi}{3}\right) = 0.5$ lagging

apparent power $= 28\ \text{V} \times 5.7\ \text{mA} = 159.6\ \text{mVA}$
true power $=$ apparent power $\times$ power factor
$$= 159.6 \times 0.5$$
$$= 79.8\ \text{mW}$$

CHAPTER SUMMARY

☐ Sinewave Relationships

$$\text{period, t (seconds)} = \frac{1}{\text{frequency, f, (Hz)}}$$

$$\text{frequency, f (Hz)} = \frac{1}{\text{period, t (seconds)}}$$

$$\text{generated frequency, f} = \frac{\text{N (rpm)} \times \text{p (number of pairs of poles)}}{60}\ \text{Hz}$$

$$\text{angular velocity, } \omega = 2\ \pi f = \frac{2\ \pi}{\text{T}}\ \text{radians per second}$$

$$\text{one cycle} = 360° = 2\ \pi\ \text{radians}$$

$$\phi\ \text{(degrees)} = \frac{180}{\pi} \times \phi\ \text{(radians)}$$

$$= 57.296 \times \phi\ \text{(radians)}$$

$$\phi\ \text{(radians)} = \frac{\pi}{180} \times \phi\ \text{(degrees)} = 0.01745 \times \phi\ \text{(degrees)}$$

☐ Sinewave equations

$$\text{instantaneous voltage, e} = E_{max} \sin \omega t = E_{max} \sin 2\pi\ \text{ft}$$
$$= E_{max} \sin \theta$$

$$E_{max}\ \text{(peak value)} = \frac{E_{p\text{-}p}}{2} = E_{rms} \times \sqrt{2} = E_{rms} \times 1.414$$

$$= E_{av} \times \frac{\pi}{2}$$

$$= E_{av} \times 1.5708$$

$$E_{rms}\ \text{(effective value)} = E_{p\text{-}p}/(2\sqrt{2}) = E_{p\text{-}p} \times 0.3535$$
$$= E_{max}/\sqrt{2} = E_{max} \times 0.707$$
$$= E_{av} \times \pi/(2\sqrt{2}) = E_{av} \times 1.11$$

$$E_{p\text{-}p}\ \text{(peak-to-peak value)} = 2\ E_{max} = 2\sqrt{2}\ E = 2.828\ E_{rms}$$
$$= E_{av} \times \pi = E_{av} \times 3.1416$$

E_{av} (average over an alternation $= E_{p\text{-}p}/\pi = E_{p\text{-}p} \times 0.318$
$$= E_{rms} \times 2\sqrt{2}/\pi = E_{rms} \times 0.9$$
$$= E_{max} \times 2/\pi = E_{max} \times 0.637$$

☐ Phase Difference

$e_1 = E_{1max} \sin \omega t$ lags $e_2 = E_{2max} \sin (\omega t + \phi)$
by ϕ radians (e_2 leads e_1 by ϕ radians)

$$e_1 + e_2 = \sqrt{E_{1max}^2 + E_{2max}^2 + 2\,E_{1max}E_{2max} \cos \phi}\ \sin (\omega t + \theta)$$

where $\theta = \text{inv tan} \left(\dfrac{E_{2max} \sin \phi}{E_{1max} + E_{2max} \cos \phi} \right)$

$$e_1 - e_2 = \sqrt{E_{1max}^2 + E_{2max}^2 - 2\,E_{1max}E_{2max} \cos \phi}\ \sin (\omega t + \theta)$$

where $\theta = \text{inv tan} \left(\dfrac{E_{2max} \sin \phi}{E_{2max} \cos \phi - E_{1max}} \right)$

☐ Resistance in ac

e and i are in phase. $\phi = 0°$

$$\frac{E_{rms}}{I_{rms}} = R\,\Omega$$

average power $= E_{rms} \times I_{rms} = \dfrac{E_{rms}^2}{R} = I_{rms}^2 \times R$ watts

☐ Inductance in ac

e leads i by $\dfrac{\pi}{2}$ radians

$$\phi = +90°, \frac{E_{rms}}{I_{rms}} = 2\,\pi f L = \omega L = X_L\,\Omega$$

average power $= 0$

reactive power $= E_{rms} \times I_{rms}$ VArs

Inductive reactances in series:

$$X_{LT} = X_{L1} + X_{L2} + X_{L3} + \cdots + X_{LN}.$$

Inductive reactances in parallel:

$$\frac{1}{X_{LT}} = \frac{1}{X_{L1}} + \frac{1}{X_{L2}} + \frac{1}{X_{L3}} + \cdots + \frac{1}{X_{LN}}$$

☐ Capacitance in ac

i leads e by $\dfrac{\pi}{2}$ radians. $\phi = -90°$

$$\frac{E_{rms}}{I_{rms}} = \frac{1}{2\,\pi f C} = \frac{1}{\omega C} = X_C\,\Omega$$

average power $= 0$

reactive power $= E_{rms} \times I_{rms}$ VArs

capacitive reactances in series:

$$X_{CT} = X_{C1} + X_{C2} + X_{C3} + \cdots + X_{CN}.$$

capacitive reactances in parallel:

$$\frac{1}{X_{CT}} = \frac{1}{X_{C1}} + \frac{1}{X_{C2}} + \frac{1}{X_{C3}} + \cdots + \frac{1}{X_{CN}}$$

☐ General Ac circuit

magnitude of impedance, $Z = E_{rms}/I_{rms}$ ohms

apparent power $= E_{rms} \times I_{rms}$ VA

$$\text{power factor} = \frac{\text{true power}}{\text{apparent power}} = \cos \phi$$

10
Alternating-Circuit Currents

IN THIS CHAPTER you will learn about the principles of alternating current circuits and in particular:

☐ How to analyze series ac circuits consisting of an inductor and a resistor, a capacitor and a resistor, an inductor and a capacitor, and an inductor, a capacitor, and a resistor.

☐ About the concepts of true power, reactive power, apparent power, and power factor.

☐ How to analyze parallel ac circuits consisting of an inductor and a resistor, a capacitor and a resistor, an inductor and a capacitor, and an inductor, a capacitor, and a resistor.

☐ How to use the properties of conductance, susceptance, and admittance to solve parallel ac circuits.

☐ The conditions required in order to obtain maximum power transfer from an ac source to a load impedance.

Referring to Fig. 10-1, voltage v_R is in phase with i, which lags v_L by 90°. The phasor sum of v_R and v_L is the supply voltage, e, which leads i (the current flowing through both R and L) by the phase angle ϕ.

SINEWAVE INPUT TO L AND R IN SERIES

The total opposition to current flow is measured by the impedance, z, which is defined as the ratio of e to i and is the sum of the resistance and the reactance phasors. The magnitude of z is

$$Z = E/I = \sqrt{(R^2 + X_L^2)} = \sqrt{(R^2 + 4\,\pi^2\,f^2\,L^2)}$$

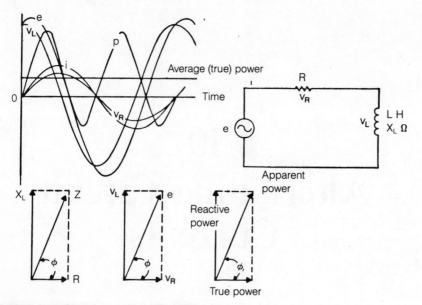

Fig. 10-1 L and R in series; X_L is greater than R.

and the rms voltages are

$$v_R = IR$$
$$v_L = IX_L$$
$$E = \sqrt{(v_R{}^2 + v_L{}^2)}$$

It follows that

$$R = \sqrt{(Z^2 - X_L{}^2)}$$
$$X_L = \sqrt{(Z^2 - R^2)}$$
$$v_R = \sqrt{(E^2 - v_L{}^2)}$$

and

$$v_L = \sqrt{(E^2 - v_R{}^2)}$$

The instantaneous power, p, is equal to e × i; its waveform is a second harmonic curve as shown in Fig. 10-1.

Only the resistance in the circuit dissipates power; therefore the apparent power in the circuit is

$$EI = \frac{E^2}{Z} = I^2Z$$

and is measured in volt-amperes (VA). The idle, wattless or reactive power associated with the inductor is

$$I^2X_L = \frac{v_L{}^2}{X_L} = Iv_L$$

and is measured in voltamperes reactive (VArs). The true power in watts is

$$I^2R = \frac{v_R^2}{R} = Iv_R$$

Then apparent power $= \sqrt{\text{true power}^2 + \text{reactive power}^2}$. This relationship can be derived from the power phasor diagram in Fig. 10-1.

The power factor is defined as the ratio of the true power to the apparent power. Therefore, the power factor, which has no units, is

$$\frac{\text{true power}}{\text{apparent power}} = \frac{I^2R}{I^2Z} = \frac{R}{Z} = \cos \phi$$

Also,

$$\sin \phi = \frac{X_L}{Z} = \frac{\text{reactive power}}{\text{apparent power}}$$

and

$$\tan \phi = \frac{X_L}{R} = \frac{\text{reactive power}}{\text{true power}}$$

The true power in watts is EI $\times$ the power factor. Note that a wattmeter used for ac measurements automatically takes into account the power factor and indicates the true power. It follows that

$$I = \frac{\text{true power}}{E \times \text{power factor}}$$

and

$$E = \frac{\text{true power}}{I \times \text{power factor}}$$

If $R = X_L$, the phase angle of the supply voltage relative to the supply current is $+45°$ and the power factor is 0.707. With ϕ a positive angle, the power factor is said to be lagging. For an entirely resistive circuit, $\phi = 0°$, and the power factor is unity; for an entirely reactive circuit, $\phi = \pm90°$, and the power factor is zero. Under no circumstances can the power factor of a circuit exceed 1.

Example 10-1

In Fig. 10-1, L = 150 μH, R = 680 Ω, and the supply voltage is 1 V, 800 kHz. Calculate the values of Z, V_L, V_R, ϕ, true power, apparent power, and power factor.

Solution

inductive reactance, $X_L = 2 \pi fL = 2 \times \pi \times 800 \times 10^3 \times 150 \times 10^{-6}$
$$= 754 \ \Omega$$

$$\text{impedance, } Z = \sqrt{R^2 + X_L{}^2} = \sqrt{680^2 + 754^2} = 1015 \ \Omega$$

$$\text{current, } I = \frac{E}{Z} = \frac{1 \text{ V}}{1015 \ \Omega} = 0.985 \text{ mA}$$

$$V_L = IX_L = 0.985 \text{ mA} \times 754 \ \Omega = 0.743 \text{ V}$$
$$V_R = IR = 0.985 \text{ mA} \times 680 \ \Omega = 0.670 \text{ V}$$

$$\text{Voltage check: } E = \sqrt{V_R{}^2 + V_L{}^2} = \sqrt{0.670^2 + 0.743^2} = 1.00 \text{ V}$$
$$\text{apparent power} = E \times I = 1 \text{ V} \times 0.985 \text{ mA} = 0.985 \text{ mVA}$$
$$\text{reactive power} = V_L \times I = 0.743 \text{ V} \times 0.985 \text{ mA} = 0.732 \text{ mVAr.}$$
$$\text{true power} = V_R \times I = 0.670 \text{ V} \times 0.985 \text{ mA} = 0.660 \text{ mW}$$

Power check:

$$\text{apparent power} = \sqrt{(\text{true power})^2 + (\text{reactive power})^2}$$
$$= \sqrt{0.660^2 + 0.732^2}$$
$$= 0.986 \text{ mVA}$$

The slight difference between 0.985 mVA and 0.986 mVA is caused by rounding off.

$$\text{power factor} = \frac{\text{true power}}{\text{apparent power}} = \frac{0.660 \text{ mW}}{0.985 \text{ mVA}} = 0.67, \text{ lagging}$$

$$\text{phase angle, } \phi = \text{inv cos } 0.67 = +47.9°$$

SINEWAVE INPUT TO C AND R IN SERIES

Referring to Fig. 10-2, the equations for C and R in series can be derived from those given earlier by substituting

$$X_C = \frac{1}{2 \pi fC} = \frac{0.159}{fC} \ \Omega$$

for X_L. The only other differences are

1. V_C lags i by 90°.
2. e lags i by the phase angle, ϕ, which now negative.
3. The power factor is considered to be leading.

Example 10-2

In Fig. 10-2, C = 0.068 μF, R = 820 Ω, and the supply voltage is 15 V, 4.5 kHz. Calculate the values of Z, V_C, V_R, ϕ, true power, apparent power, and power factor.

Solution

$$\text{capacitive reactance, } X_C = \frac{1}{2 \pi fC}$$

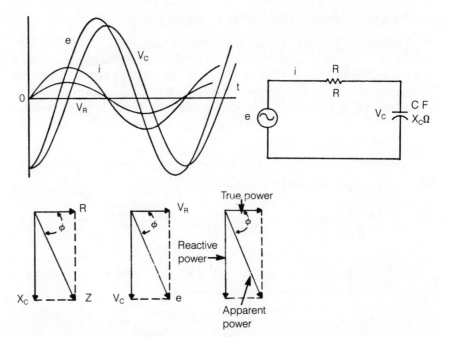

Fig. 10-2 C and R in series; X_C is greater than R.

$$= \frac{1}{2 \times \pi \times 4.5 \times 10^3 \times 0.068 \times 10^{-6}}$$
$$= 520 \ \Omega$$

$$\text{impedance, } Z = \sqrt{R^2 + X_C^2} = \sqrt{820^2 + 520^2} = 971 \ \Omega$$

$$\text{current, } I = \frac{E}{Z} = \frac{15 \text{ V}}{971 \ \Omega} = 15.45 \text{ mA}$$

$$V_C = IX_C = 15.45 \text{ mA} \times 520 \ \Omega = 8.03 \text{ V}$$
$$V_R = IR = 15.45 \text{ mA} \times 820 \ \Omega = 12.67 \text{ V}$$

Voltage check: $E = \sqrt{V_C^2 + V_R^2} = \sqrt{8.03^2 + 12.67^2} = 15.00$ V
apparent power $= E \times I = 15 \text{ V} \times 15.45 \text{ mA} = 231.75$ mVA
reactive power $= V_C \times I = 8.03 \times 15.45 \text{ mA} = 124.06$ mVAr
true power $= V_R \times I = 12.67 \text{ V} \times 15.45 \text{ mA} = 195.75$ mW

Power check:

$$\text{apparent power} = \sqrt{(\text{true power})^2 + (\text{reactive power})^2}$$
$$= \sqrt{195.75^2 + 124.06^2}$$
$$= 231.75 \text{ mVA}$$

$$\text{power factor} = \frac{\text{true power}}{\text{apparent power}} = \frac{195.75}{231.75} = 0.845, \text{ leading}$$

$$\text{phase angle, } \phi = \text{inv cos } 0.845 = -32.4°$$

SINEWAVE INPUT TO L AND C IN SERIES

Referring to Fig. 10-3, v_L leads i by 90°. Therefore v_L and v_C are 180° out of phase, and when added together to produce their phasor sum, e, their magnitudes, v_L and v_C, tend to cancel. If X_L is greater than X_C, the total impedance, $Z = X_L - X_C$, the source voltage, $E = v_L - v_C$, and e leads i by 90° ($\phi = +90°$); the circuit behaves inductively. If X_L is less than X_C, $Z = X_C - X_L$, $E = v_C - v_L$ and e lags i by 90° ($\phi = -90°$); the circuit behaves capacitively. In both cases, the circuit's power factor is zero (it is

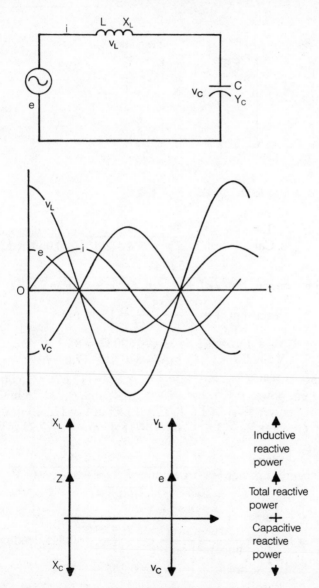

Fig. 10-3 L and C in series; X_L is greater than WC.

assumed that there are no losses associated with either the inductor or the capacitor). If $X_L = X_C$, $v_L = v_C$, $Z = 0$, I is theoretically infinite, and the series combination of L and C therefore behaves as a short circuit ($\phi = 0°$). For particular values of L and C, this situation occurs at a frequency

$$f = \frac{1}{2\,\pi\,\sqrt{LC}} = \frac{0.159}{\sqrt{LC}}$$

where

> f is in Hz
> L in henrys
> C in farads

Then

$$L = \frac{1}{4\,\pi^2 f^2 C} = \frac{0.0253}{f^2 C}\,H$$

and

$$C = \frac{1}{4\,\pi^2 f^2 L} = \frac{0.0253}{f^2 L}\,F$$

Example 10-3

In Fig. 10-3, $L = 125\ \mu H$, $C = 275$ pF, and the source voltage is 100 mV, 750 kHz. Calculate the values of Z, I, v_L, v_C, and the total reactive power.

Solution

inductive reactance, $X_L = 2\,\pi f L$
$$= 2 \times \pi \times 750 \times 10^3 \times 125 \times 10^{-6}$$
$$= 589\ \Omega$$

capacitance reactance, $X_C = \dfrac{1}{2\,\pi f C}$

$$= \frac{1}{2 \times \pi \times 750 \times 10^3 \times 275 \times 10^{-12}}$$
$$= 772\ \Omega$$

impedance, $Z = X_C - X_L = 772 - 589 = 183\ \Omega$

and is capacitive.

$$\text{current, } I = \frac{E}{Z} = \frac{100\ mV}{183\ \Omega} = 0.546\ mA$$

$$v_L = I \times X_L = 0.546\ mA \times 589\ \Omega = 322\ mV$$
$$v_C = I \times X_C = 0.546\ mA \times 772\ \Omega = 422\ mV$$

Voltage check: source voltage, $E = v_C - v_L = 422 - 322 = 100$ mV.

Notice that v_L and v_C are each greater than the source voltage.

$$\text{inductive reactive power} = I \times v_L = 0.546 \text{ mA} \times 322 \text{ mV}$$
$$= 176 \ \mu\text{VAr}$$
$$\text{capacitive reactive power} = I \times v_C = 0.546 \text{ mA} \times 422 \text{ mV}$$
$$= 230 \ \mu\text{VAr}$$
$$\text{total reactive power} = 230 - 176 = 54 \ \mu\text{VAr}$$
$$\text{Power check: total reactive power} = E \times I = 100 \text{ mV} \times 0.546 \text{ mA}$$
$$= 54.6 \ \mu\text{VAr}$$

The power factor is zero, leading, and the phase angle is $-90°$.

SINEWAVE INPUT TO L, C, AND R IN SERIES

Referring to Fig. 10-4, v_R is in phase with i, v_L leads i by 90° and v_C lags i by 90°. Assuming that X_L is greater than X_C, the phasor sum, v_X, of v_L and v_C is in phase with v_L while e (which is the phasor sum of v_L, v_C, and v_R) leads i by the phase angle ϕ The circuit, therefore, behaves inductively and is positive. If X_C is greater than X_L, the phasor sum of v_L and v_C is in phase with v_C, e lags i, the circuit is capacitive, and ϕ is negative.

Note that the special case of $X_L = X_C$ is concerned with the phenomenon of resonance, which is covered in Chapter 11.

The magnitude of the impedance is $Z = \sqrt{(R^2 + X^2)}$ where X is the net reactance that is the difference in value between X_L and X_C.

The supply voltage is $E = \sqrt{(v_R{}^2 + v_X{}^2)}$, where v_X is the difference in value between v_L and v_C. The current is $I = E/Z$, and $v_R = IR$, $v_L = IX_L$, $v_C = IX_C$.

Note that although v_R cannot exceed the supply voltage, E, either v_L or v_C (or both v_L and v_C) can be greater than E.

$$\text{apparent power} = EI = I^2Z = E^2/Z \text{ volt-amperes.}$$
$$\text{true power in watts} = I^2R = v_R{}^2/R = Iv_R = EI \times \text{power factor}$$

$$\text{power factor} = \frac{\text{true power}}{\text{apparent power}} = \cos \phi = R/Z$$

and

$$\text{Sin } \phi = \frac{X}{Z}, \text{ tan } \phi = \frac{X}{R}$$

If $R = X$, the phase angle is 45°.

Example 10-4

In Fig. 10-4, L = 2.5 H, C = 5 μF, and R = 350 Ω. If the source voltage is 110 V, 60 Hz, calculate the values of the total impedance, v_L, v_C, v_R, the power factor, and the phase angle.

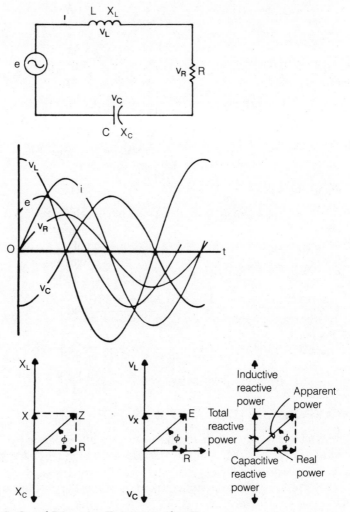

Fig. 10-4 *L, C, and R in series; X_L is greater than X_C.*

Solution

$$\text{inductive reactance, } X_L = 2\,\pi fL = 2 \times \pi \times 60 \times 2.5$$
$$= 942.5\ \Omega$$

$$\text{capacitive reactance, } X_C = \frac{1}{2\,\pi fC}$$

$$= \frac{1}{2 \times \pi \times 60 \times 5 \times 10^{-6}}$$

$$= 530.5\ \Omega$$

$$\text{total impedance, } Z = \sqrt{R^2 + (X_L - X_C)^2}$$
$$= \sqrt{350^2 + (942.5 - 530.5)^2}$$
$$= 541 \ \Omega$$

Because X_L is greater than X_C, the impedance is inductive.

$$\text{current, } I = \frac{E}{Z} = \frac{110 \text{ V}}{541 \ \Omega} = 0.204 \text{ A}$$

$$V_L = I \times X_L = 0.204 \text{ A} \times 942.5 \ \Omega = 191.8 \text{ V}$$
$$V_C = I \times X_C = 0.204 \text{ A} \times 530.5 \ \Omega = 108.2 \text{ V}$$
$$V_R = I \times R = 0.204 \text{ A} \times 350 \ \Omega = 71.4 \text{ V}$$

Voltage check:

$$\text{source voltage, } E = \sqrt{71.4^2 + (191.8 - 108.2)^2}$$
$$= \sqrt{71.4 + 83.6^2} = 110 \text{ V}$$
$$\text{true power} = I \times V_R = 0.204 \text{ A} \times 71.4 \text{ V} = 14.6 \text{ W}$$
$$\text{reactive power} = I \times (V_L - V_C) = 0.204 \text{ A} \times 83.6 \text{ V}$$
$$= 17.1 \text{ VAr}$$
$$\text{apparent power} = I \times E = 0.204 \text{ A} \times 110 \text{ V} = 22.4 \text{ VA}$$

Power check: apparent power $= \sqrt{14.6^2 + 17.1^2} = 22.5$ VA

$$\text{power factor} = \frac{\text{true power}}{\text{apparent power}} = \frac{14.6}{22.5} = 0.65, \text{ lagging}$$

Phase angle, $\phi = \text{inv cos} \cdot 0.65 = +49.5°$.

SINEWAVE INPUT TO L AND R IN PARALLEL

Referring to Fig. 10-5, i_R is in phase with e while i_L lags e by 90°. The phasor sum of i_L and i_R is the supply current, i_S, which lags e by the phase angle, ϕ.

The magnitude of the total impedance is given by the product-over-sum formula

$$Z_T = \frac{R X_L}{\sqrt{R^2 + X_L^2}}$$

Note that Z_T is smaller in value than either R or X_L.
The supply current is

$$I_S = \frac{E}{Z_T} = \sqrt{(I_R^2 + I_L^2)}$$

The lagging power factor is

$$\cos \phi = \frac{Z_T}{R} \left(not \ \frac{R}{Z_T} \right)$$

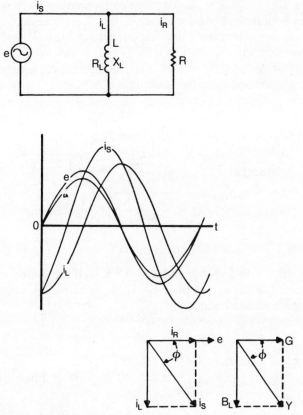

Fig. 10-5 L and R in parallel.

The apparent power is

$$EI_S \text{ volt-amperes}$$

The true power is

$$EI_S \times \text{power factor} = I_R{}^2R = \frac{E^2}{R} \text{ watts}$$

Also

$$\sin \phi = \frac{Z_T}{X_L}$$

and

$$\tan \phi = \frac{R}{X}$$

For parallel circuits, it can be more convenient to work with conductance G, susceptance B, and admittance Y, which are, respectively, the reciprocals of resistance, reactance and impedance, and are all measured in siemens. Therefore the conductance is $G = 1/R$ (as in DC circuits), the inductive susceptance, is $B_L = 1/X_L$ and the admittance, which is the phasor sum of the conductance and the susceptance, is $Y_T = 1/Z_T$. In terms of their magnitudes,

$$\sin \phi = \frac{B_L}{Y}, \cos \phi = \frac{G}{Y}, \tan \phi = \frac{B_L}{G}$$

$$\text{admittance, } Y_T = \sqrt{G^2 + B_L^2} = \sqrt{\frac{1}{R^2} + \frac{1}{X_L^2}}$$

$$= \frac{\sqrt{R^2 + X_L^2}}{RX_L} = \frac{1}{Z_T} \text{ siemens}$$

Example 10-5

In Fig. 10-5, $L = 550$ mH and $R = 6.8$ kΩ. If the source voltage is 8 V, 2.5 kHz, calculate the values of Z, I_R, I_L, I_S, true power, apparent power, power factor, and phase angle.

Solution

$$\text{inductive reactance, } X_L = 2 \pi fL$$
$$= 2 \times \pi \times 2.5 \times 10^3 \times 550 \times 10^{-3} \ \Omega$$
$$= 8.64 \text{ k}\Omega$$

There are two ways to calculate the total impedance.

Method 1

$$\text{total impedance, } Z_T = \frac{R \times X_L}{\sqrt{R^2 + X_L^2}} = \frac{6.8 \times 8.64}{\sqrt{6.8^2 + 8.64^2}}$$
$$= 5.34 \text{ k}\Omega$$

Method 2

$$I_R = \frac{8 \text{ V}}{6.8 \text{ k}\Omega} = 1.18 \text{ mA}$$

$$I_L = \frac{8 \text{ V}}{8.64 \text{ k}\Omega} = 0.926 \text{ mA}$$

$$I_S = \sqrt{1.18^2 + 0.926^2} = 1.5 \text{ mA}$$

$$\text{total impedance, } Z_T = \frac{E}{I_S} = \frac{8 \text{ V}}{1.5 \text{ mA}} = 5.33 \text{ k}\Omega$$

If no source voltage is given, method 2 can be used by assuming any convenient value for E.

Method 3

$$\text{inductive susceptance, } B_L = \frac{1}{8.64 \text{ k}\Omega} = 1.16 \times 10^{-4} \text{ siemens}$$

$$\text{conductance, } G = \frac{1}{6.8 \text{ k}\Omega} = 1.47 \times 10^{-4} \text{ siemens}$$

$$\text{admittance, } Y_T = \sqrt{G^2 + B_L^2}$$
$$= \sqrt{(1.47 \times 10^{-4})^2 + (1.16 \times 10^{-4})^2}$$
$$= 1.87 \times 10^{-4} \text{ siemens}$$

$$\text{impedance, } Z_T = \frac{1}{Y_T} = \frac{1}{1.87 \times 10^{-4}} = 5.34 \text{ k}\Omega$$

$$\text{true power} = \frac{E^2}{R} = \frac{(8 \text{ V})^2}{6.8 \text{ k}\Omega} = 9.41 \text{ mW}$$

Note that the true power is constant and independent of frequency.

$$\text{reactive power} = \frac{E^2}{X_L} = \frac{(8 \text{ V})^2}{8.64 \text{ k}\Omega} = 7.41 \text{ mVAr}$$

$$\text{apparent power} = \frac{E^2}{Z} = \frac{(8 \text{ V})^2}{5.34 \text{ k}\Omega} = 11.99 \text{ mVA}$$

Power check:

$$\text{apparent power} = \sqrt{(\text{true power})^2 + (\text{reactive power})^2}$$
$$= \sqrt{9.41^2 + 7.41^2}$$
$$= 11.98 \text{ mVA}$$

$$\text{power factor} = \frac{\text{true power}}{\text{apparent power}} = \frac{9.41}{11.99} = 0.8$$

$$= \frac{Z_T}{R} = \frac{5.34}{6.8} = 0.8, \text{ lagging}$$

$$\text{phase angle, } \phi = \text{inv cos } 0.8 = +38°$$

SINEWAVE INPUT TO C AND R IN PARALLEL

The equations for C and R in parallel can be derived from those given earlier by substituting X_C for X_L, I_C for I_L, and B_C for B_L. Referring to Fig. 10-6, the only other differences are

1. i_C lags e by 90°.
2. i_S leads e by the phase angle, ϕ, which is negative.
3. The power factor is now leading.

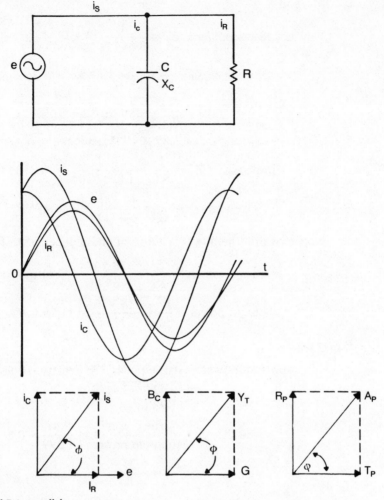

Fig. 10-6 C and R in parallel.

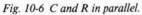

Example 10-6

In Fig. 10-6, C = 35 pF and R = 3.3 kΩ. If the source voltage 50 mV, 1.8 MHz, calculate the values of the total impedance, I_R, I_C, I_S, true power, power factor, and phase angle.

Solution

$$\text{capacitive reactance, } X_C = \frac{1}{2\pi f C}$$

$$= \frac{1}{2 \times \pi \times 1.8 \times 10^6 \times 35 \times 10^{-12}}$$

$$= 2.53 \text{ k}\Omega$$

$$\text{total impedance, } Z_T = \frac{R \times X_C}{\sqrt{R^2 + X_C^2}} = \frac{3.3 \times 2.53}{\sqrt{3.3^2 + 2.53^2}}$$
$$= 2.0 \text{ k}\Omega$$

$$I_R = \frac{50 \text{ mV}}{3.3 \text{ k}\Omega} = 15.2 \text{ } \mu A$$

$$I_C = \frac{50 \text{ mV}}{2.53 \text{ k}\Omega} = 19.8 \text{ } \mu A$$

$$\text{supply current, } I_S = \sqrt{I_R^2 + I_C^2} = \sqrt{15.2^2 + 19.8^2}$$
$$= 25.0 \text{ } \mu A$$

Check:

$$\text{total impedance} = \frac{50 \text{ mV}}{25.0 \text{ } \mu A} = 2.0 \text{ k}\Omega$$

$$\text{true power} = \frac{E^2}{R} = \frac{(50 \text{ mV})^2}{3.3 \text{ k}\Omega} = 0.758 \text{ } \mu W$$

$$\text{reactive power} = \frac{E^2}{X_C} = \frac{(50 \text{ mV})^2}{2.53 \text{ k}\Omega} = 0.988 \text{ } \mu VAr$$

$$\text{apparent power} = E \times I_S = 50 \text{ mV} \times 25 \text{ } \mu A = 1.25 \text{ } \mu VA$$

Power check:

$$\text{apparent power} = \sqrt{(\text{true power})^2 + (\text{reactive power})^2}$$
$$= \sqrt{0.758^2 + 0.988^2}$$
$$= 1.25 \text{ } \mu VA$$

$$\text{power factor} = \frac{\text{true power}}{\text{apparent power}} = \frac{0.758}{1.25} = 0.61, \text{ leading}$$

$$\text{phase angle, } \phi = \text{inv cos } 0.61 = -53°$$

SINEWAVE INPUT TO L AND C IN PARALLEL

Referring to Fig. 10-7, i_C leads e by 90° while i_L lags e by 90°; i_L and i_C are therefore 180° out of phase. Consequently, when i_L and i_C are added together to produce their phasor sum i_S, their magnitudes I_L and I_C tend to cancel.

If X_L is greater than X_C, then I_L ($= E/X_L$) is less than I_C ($= E/X_C$), and $I_S = I_C - I_L$. Supply current i_S leads e by 90° ($\phi = -90°$) and the circuit behaves capacitively. This is the opposite result from the equivalent series case, in which if X_L is greater than X_C, the circuit behaves inductively.

If X_L is less than X_C, then I_L is greater than I_C and $I_S = I_L - I_C$. Supply current i_S lags e by 90° ($\phi = +90°$), and the circuit is inductive.

Whether X_L is greater or less than X_C, the power factor in either case is zero (assuming that there are no resistance losses associated with either L or C).

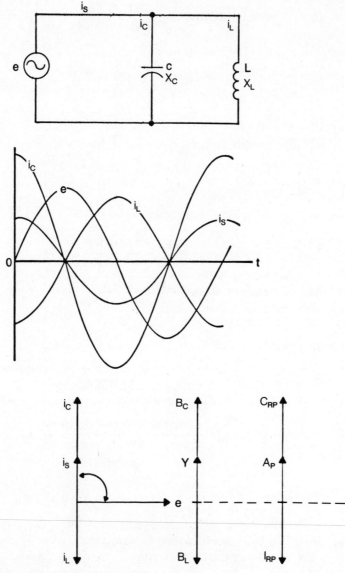

Fig. 10-7 L and C in parallel; X_L is greater than X_C.

The total impedance of the circuit, Z_T is $X_L \times X_C/X$, where X is the difference in value between X_L and X_C. Note that the magnitude of Z is always greater than that of either X_L or X_C, and can be greater than both X_L and X_C.

The total admittance of the circuit, $Y_T = B$, where B is the difference in value between $B_L = 1/X_L$ and $B_C = 1/X_C$.

If $X_L = X_C$, $I_L = I_C$, $I_S = 0$, Z_T is theoretically infinite and Y is therefore zero. The parallel combination of L and C then behaves as an open circuit. This situation will also occur at the frequency:

$$f = \frac{1}{2\,\pi\,\sqrt{LC}} = \frac{0.159}{\sqrt{LC}}\ \text{Hz}$$

Example 10-7

In Fig. 10-7, L = 75 μH and C = 315 pF. If the source voltage is 20 mV, 775 kHz, calculate the value of the total reactance, I_L, I_C, I_S, and the total reactive power.

Solution

inductive reactance, $X_L = 2\ \pi fL$
$$= 2 \times \pi \times 775 \times 10^3 \times 75 \times 10^{-6}$$
$$= 365\ \Omega$$

capacitive reactance, $X_C = \dfrac{1}{2\ \pi\ fC}$

$$= \frac{1}{2 \times \pi \times 775 \times 10^3 \times 315 \times 10^{-12}}$$
$$= 652\ \Omega$$

There are two methods for finding the total reactance.

Method 1

$$\text{total reactance, } X = \frac{X_L \times X_C}{X_C - X_L} = \frac{365 \times 652}{652 - 365}$$

$$= \frac{365 \times 652}{287}$$

$$= 830\ \Omega,\ \text{rounded off}$$

Method 2

$$I_L = \frac{E}{X_L} = \frac{20\ \text{mV}}{365\ \Omega} = 0.0548\ \text{mA}$$

$$I_C = \frac{E}{X_C} = \frac{20\ \text{mV}}{652\ \Omega} = 0.0307\ \text{mA}$$

$$I_S = I_L - I_C = 0.0548 - 0.0307 = 0.0241\ \text{mA}$$

Because I_L is greater than I_C, the circuit is inductive.

$$\text{total reactance, } X = \frac{E}{I_S} = \frac{20\ \text{mV}}{0.0241\ \text{mA}} = 830\ \Omega,\ \text{rounded off}$$

Method 3

$$\text{inductive susceptance, } B_L = \frac{1}{X_L} = \frac{1}{365}$$

$$= 2.74 \times 10^{-3} \text{ siemens}$$

$$\text{capacitive susceptance, } B_C = \frac{1}{X_C} = \frac{1}{652}$$

$$= 1.53 \times 10^{-3} \text{ siemens}$$

$$\text{total susceptance, } B = B_L - B_C = 1.21 \times 10^{-3} \text{ siemens}$$

$$\text{total reactance, } X = \frac{1}{B} = \frac{1}{1.21 \times 10^{-3}} = 830 \ \Omega, \text{ rounded off}$$

$$\text{total reactive power} = E \times I_S = 10 \text{ mV} \times 0.0241 \text{ mA}$$

$$= 0.48 \ \mu\text{VAr, rounded off}$$

The power factor is zero, lagging, and the phase angle, ϕ, is $+90°$.

SINEWAVE INPUT TO L, C, AND R IN PARALLEL

Referring to Fig. 10-8, current i_R is in phase with e, i_C leads e by 90°, and i_L lags e by 90°; i_L and i_C are therefore 180° out of phase so that when they are added together to produce their phasor sum i_X, their magnitudes, I_L and I_C, tend to cancel. The phasor sum of i_R and i_X is the supply current, i_S. If X_L is greater than X_C, I_L is less than I_C, and i_X will be in phase with i_C. Supply current i_S leads e by the phase angle, ϕ, and the circuit is capacitive. This is the opposite result from the series L, C, R circuit, which is inductive when X_L is greater than X_C. If X_L is less than X_C, i_X is in phase with i_L. Supply current i_S lags e, and the circuit is inductive. The special case of $X_L = X_C$ will be covered in the Chapter 11 discussion on resonance.

The total reactance of L and C in parallel is

$$X_T = \frac{X_L \times X_C}{X}$$

where X is the difference in value between X_L and X_C. The total impedance is

$$Z_T = \frac{RX_T}{\sqrt{R^2 + X_T{}^2}}$$

which is always less than the value of R; however, Z_T can be greater than either X_L or X_C (or both X_L and X_C). The supply current is

$$I_S = \frac{E}{Z} = \sqrt{I_R{}^2 + I_X{}^2}$$

where I_X is the difference in value between I_L and I_C.

The true power is

$$\frac{E^2}{R} = I_R{}^2 R = EI_R \text{ watts}$$

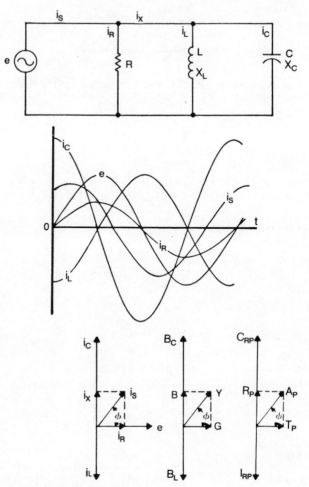

Fig. 10-8 *L, C, and R in parallel; X_L is greater than X_C.*

The apparent power is

$$\frac{E^2}{Z} = I_S^2 Z = EI_S \text{ volt-amperes}$$

The power factor is

$$\frac{Z_T}{R} \left(\text{not } \frac{R}{Z}\right)$$

so that

$$\text{Cos } \phi = \frac{Z_T}{R}$$

and

$$\sin \phi = \frac{Z}{X_T}, \ \tan \phi = \frac{R}{X_T}$$

The total susceptance, B, of L and C in parallel is the difference in value between

$$B_L = \frac{1}{X_L}$$

and

$$B_C = \frac{1}{X_C}$$

Total admittance is

$$Y_T = \frac{1}{Z_T} = \sqrt{(G^2 + B^2)} \text{ siemens}$$

where the conductance is

$$G = \frac{1}{R}$$

Then

$$\cos \phi = \frac{G}{Y}, \ \sin \phi = \frac{B}{Y}, \ \tan \phi = \frac{B}{G}$$

Example 10-8

In Fig. 10-8, $R = 2.2 \text{ k}\Omega$, $L = 3.5 \ \mu\text{H}$, $C = 8.5 \text{ pF}$. If the source voltage is 200 mV, 35 MHz, calculate the values of the total impedance, I_R, I_L, I_C, I_S, true power, apparent power, power factor, and phase angle.

Solution

inductive reactance, $X_L = 2 \pi f L$
$$= 2 \times \pi \times 35 \times 10^6 \times 3.5 \times 10^{-6}$$
$$= 770 \ \Omega$$

capacitive reactance, $X_C = \dfrac{1}{2 \pi f C}$

$$= \frac{1}{2 \times \pi \times 35 \times 10^6 \times 8.5 \times 10^{-12}}$$
$$= 535 \ \Omega$$

total reactance, $X_T = \dfrac{X_L \times X_C}{X_L - X_C} = \dfrac{770 \times 535}{770 - 535}$

$$= \frac{770 \times 535}{235} \,\Omega = 1.75 \text{ k}\Omega$$

$$\text{total impedance} = \frac{2.2 \times 1.75}{\sqrt{2.2^2 + 1.75^2}} = 1.37 \text{ k}\Omega$$

$$I_R = \frac{200 \text{ mV}}{2.2 \text{ k}\Omega} = 90.91 \,\mu A$$

$$I_L = \frac{200 \text{ mV}}{770 \,\Omega} = 259.7 \,\mu A$$

$$I_C = \frac{200 \text{ mV}}{535 \,\Omega} = 373.8 \,\mu A$$

$$\text{supply current, } I_S = \sqrt{90.91^2 + (373.8 - 259.7)^2}$$
$$= \sqrt{90.91^2 + 114.1^2}$$
$$= 145.9 \,\mu A$$

Because I_C is greater than I_L, the line current is capacitive.
Check:

$$\text{total impedance, } Z_T = \frac{E}{I_S} = \frac{200 \text{ mV}}{145.9 \,\mu A} = 1.37 \text{ k}\Omega$$

$$\text{true power} = \frac{E^2}{R} = \frac{(200 \text{ mV})^2}{2.2 \text{ k}\Omega} = 18.2 \,\mu W$$

$$\text{apparent power} = E \times I = 200 \text{ mV} \times 149.9 \,\mu A$$
$$= 30.0 \,\mu VA$$

$$\text{power factor} = \frac{\text{true power}}{\text{apparent power}} = \frac{18.2}{30.0}$$
$$= 0.61, \text{ rounded off}$$

Check:

$$\text{power factor} = \frac{Z_T}{R} = \frac{1.37}{2.2} = 0.62, \text{ leading}$$

$$\text{phase angle, } \phi = \text{inv cos } 0.62 = -52°$$

MAXIMUM POWER TRANSFER-THE AC CASE

With a dc voltage source, there is maximum transfer power to the load when the load resistance is matched (made equal) to the internal resistance of the source. An ac source can possess an internal impedance consisting of resistance and either inductive or capacitive reactance, while the load can be purely resistive or a combination of resistance and reactance. I discuss these two possibilities for the load separately.

Resistive Load

The magnitude of the source's total internal impedance is $\sqrt{R_S^2 + X_S^2}$, where X_S can represent either inductive or capacitive reactance (Fig. 10-9).

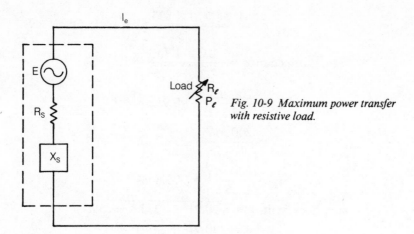

Fig. 10-9 Maximum power transfer with resistive load.

The magnitude of the circuit's total impedance is

$$Z_T = \sqrt{(R_S + R_\ell)^2 + X_S^2}.$$

$$\text{load current, } I_\ell = \frac{E}{Z_T} = \frac{E}{\sqrt{(R_S + R_\ell)^2 + X_S^2}}$$

$$\text{load power, } P_\ell = I_\ell^2 \times R_\ell = \frac{E^2 \times R_\ell}{(R_S + R_\ell)^2 + X_S^2}$$

If R_ℓ is varied, it can be shown that the load power, P_ℓ, is a maximum when $R_\ell = \sqrt{R_S^2 + X_S^2}$, or when the load resistance is matched to the magnitude of the source's internal impedance. The value of the maximum power is

$$P_{\ell(max)} = \frac{E^2}{2(R_\ell + R_S)} \text{ watts.}$$

Impedance Load

The magnitude of the total impedance of the circuit (Fig. 10-10) is

$$Z_T = \sqrt{(R_\ell + R_S)^2 + (X_\ell + X_S)^2}.$$

where X_ℓ and X_S can be either inductive or capacitive. The load current is

$$I_\ell = \frac{E}{Z_T} = \frac{E}{\sqrt{(R_\ell + R_S)^2 + (X_\ell + X_S)^2}}$$

The power in the load is

$$P_\ell = I^2 R_\ell = \frac{E^2 R_\ell}{(R_\ell + R_S)^2 + (X_\ell + X_S)^2}.$$

If X_ℓ is varied, P_ℓ has a maximum value when $X_\ell + X_S = 0$, or $X_\ell = -X_S$. X_ℓ then has the same magnitude as X_S but is opposite in nature so that if, for example X_S is inductive, X_ℓ is made capacitive, and vice-versa.

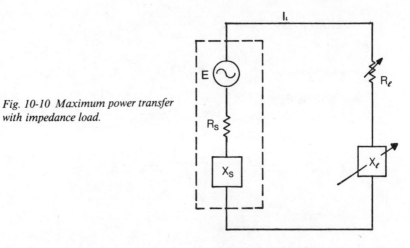

Fig. 10-10 Maximum power transfer
with impedance load.

After the reactance has been cancelled in the circuit,

$$\text{load power, } P_\ell = \frac{E^2 R_\ell}{(R_\ell + R_S)^2}.$$

P_ℓ has a maximum value when R_ℓ is matched to R_S. The value of the maximum power is

$$P_{\ell(max)} = \frac{E^2}{4 R_s}$$

and the corresponding load current is

$$I_\ell = \frac{E}{2R_s}$$

With $R_\ell = R_S$ and $X_\ell = -X_S$, the load impedance and the source's internal impedance are said to be *conjugates*.

Example 10-9

In Fig. 10-9, $R_S = 3$ kΩ, $X_S = 4$ kΩ, and $E = 12$ V. Determine the load power when R_L is (a) 3 kΩ and (b) 7 kΩ. What is the value of R_L for which the load power is a maximum? Calculate the amount of the maximum load power.

Solution

(a) total impedance, $Z_T = \sqrt{(3 + 3)^2 + 4^2}$
$$= \sqrt{52} = 7.21 \text{ k}\Omega$$

$$\text{load current, } I_\ell = \frac{E}{Z_T} = \frac{12 \text{ V}}{7.21 \text{ k}\Omega} = 1.66 \text{ mA}$$

$$\text{load power, } P_\ell = I_\ell^2 \times R_\ell = (1.66 \text{ mA})^2 \times 3 \text{ k}\Omega$$
$$= 8.3 \text{ mW}$$

(b) total impedance, $Z_T = \sqrt{(3+7)^2 + 4^2} = \sqrt{116}$
$$= 10.8 \text{ k}\Omega$$

load current, $I_\ell = \dfrac{12 \text{ V}}{10.8 \text{ k}\Omega} = 1.11 \text{ mA}$

load power, $P_\ell = I_\ell^2 \times R_\ell = (1.11 \text{ mA})^2 \times 7 \text{ k}\Omega$
$$= 8.68 \text{ mW}$$

load power is a maximum when $R_\ell = \sqrt{3^2 + 4^2}$
$$= 5 \text{ k}\Omega$$

load impedance, $Z_T = \sqrt{(3+5)^2 + 4^2} = \sqrt{80}$
$$= 8.94 \text{ k}\Omega$$

load current, $I_\ell = \dfrac{12 \text{ V}}{8.94 \text{ k}\Omega} = 1.34 \text{ mA}$

maximum load power, $P_{\ell(max)} = (1.34 \text{ mA})^2 \times 5 \text{ k}\Omega$
$$= 8.98 \text{ mW}$$

Example 10-10

In Fig. 10-10, $E = 18$ V and $R_S = 4$ kΩ. The source reactance, X_S, is capacitive and equal to 5 kΩ. Determine the values of R_ℓ and X_ℓ that will allow maximum power transfer to the load, and find the amount of the maximum load power.

Solution

For maximum power transfer to the load, the load impedance must be the conjugate of the source impedance.

Therefore $R_\ell = 4$ kΩ and X_ℓ is an inductive reactance of 5 kΩ.

$$\text{maximum load power} = \frac{E^2}{4 R_\ell} = \frac{(18 \text{ V})^2}{4 \times 4 \text{ k}\Omega} = 20.25 \text{ mW}$$

CHAPTER SUMMARY

☐ L and R in Series

$$Z = \sqrt{R^2 + X_L^2} \ \Omega$$
$$E = IZ, \ V_R = IR, \ V_L = IX_L$$
$$E = \sqrt{V_R^2 + V_L^2}$$

$$\text{true power} = I^2R = IV_R = \frac{V_R^2}{R} = E \times I \times \text{p.f watts}$$

$$\text{reactive power} = I^2X_L = IV_L = \frac{V_L}{X_L} \text{ VAr}$$

$$\text{apparent power} = I^2Z = IE = \frac{E^2}{Z} \text{ VA}$$

$$\text{apparent power} = \sqrt{(\text{true power})^2 + (\text{reactive power})^2}$$

$$\text{power factor, (p.f)} = \frac{R}{Z} = \frac{V_R}{E} = \cos \phi, \text{ lagging}$$

$\phi = \text{inv cos (power factor) and is positive because e leads i}$

$$\sin \phi = \frac{X_L}{Z} = \frac{V_L}{E}, \tan \phi = \frac{X_L}{R} = \frac{V_L}{V_R}$$

☐ C and R in Series

$$Z = \sqrt{R^2 + X_C^2} \ \Omega$$
$$E = IZ, \ V_R = IR, \ V_C = IX_C$$
$$E = \sqrt{V_R^2 + V_C^2}$$

$$\text{true power} = I^2R = IV_R = \frac{V_R^2}{R} = E \times I \times \text{p.f. watts}$$

$$\text{reactive power} = I^2X_C = IV_C = \frac{V_C^2}{X_C} \text{ VAr}$$

$$\text{apparent power} = I^2Z = IE = \frac{E^2}{Z} \text{ VA}$$

$$\text{apparent power} = \sqrt{(\text{true power})^2 + (\text{reactive power})^2}$$

$$\text{power factor} = \frac{R}{Z} = \frac{V_R}{E} = \cos \phi, \text{ leading}$$

$\phi = \text{inv cos (power factor). The sign of } \phi \text{ is negative since e lags i.}$

$$\sin \phi = \frac{X_C}{Z} = \frac{V_C}{E}$$

$$\tan \phi = \frac{X_C}{R} = \frac{V_C}{V_R}$$

☐ L and C in Series

$$Z = X_L \sim X_C$$
$$E = IZ, \ V_L = IX_L, \ V_C = IX_C$$
$$E = V_L \sim V_C$$
$$\text{true power} = 0 \text{ watts}$$
$$\text{reactive power} = I^2 (X_L \sim X_C) = EI \text{ VAr}$$
power factor = 0 and is leading if X_C is greater than X_L or lagging if X_L is greater than X_C

ϕ equals $+90°$ if X_L is greater than X_C but is $-90°$ if X_C is greater than X_L

$$X_L = X_C \text{ when the frequency, } f = \frac{1}{2 \pi \sqrt{LC}} \text{ Hz}$$

☐ L, C and R in Series

$$Z = \sqrt{R^2 + (X_L \sim X_C)^2} \; \Omega$$
$$E = IZ, \; V_R = IR, \; V_L = IX_L, \; V_C = IX_C$$
$$E = \sqrt{V_R^2 + (V_L \sim V_C)^2}$$

$$\text{true power} = I^2R = IV_R = \frac{V_R^2}{R} = E \times I \times \text{p.f. watts}$$

$$\text{reactive power} = I^2 (X_L \sim X_C) = I (V_L \sim V_C) = \frac{(V_L \sim V_C)^2}{X_L \sim X_C} \; \text{VAr}$$

$$\text{apparent power} = I^2Z = IE = \frac{E^2}{Z} \; \text{VA}$$

$$\text{apparent power} = \sqrt{(\text{true power})^2 + (\text{reactive power})^2}$$

$$\text{power factor} = \frac{R}{Z} = \frac{V_R}{E} = \cos \phi.$$

The power factor is leading if X_C is greater than X_L, but lagging if X_L is greater than X_C. $\phi = \text{inv} \cos (\text{power factor})$.

The sign of ϕ is negative if X_C is greater than X_L but positive if X_L is greater than X_C.

$$\sin \phi = \frac{X_L \sim X_C}{Z}, \; \tan \phi = \frac{X_L \sim X_C}{R}$$

☐ L and R in Parallel

$$Z_T = \frac{R \times X_L}{\sqrt{R^2 + X_L^2}} \; \Omega$$

$$I_S = \frac{E}{Z}, \; I_R = \frac{E}{R}, \; I_L = \frac{E}{X_L}$$

$$I_S = \sqrt{I_R^2 + I_L^2}$$

$$\text{true power} = \frac{E^2}{R} = I_R^2 \times R = E \times I_R = E \times I_S \times \text{p.f. watts}$$

$$\text{reactive power} = \frac{E^2}{X_L} = I_L^2 \times X_L = E \times I_L \; \text{VArs}$$

$$\text{apparent power} = \frac{E^2}{Z} = I_S^2 \times Z = E \times I_S \; \text{VA}$$

$$\text{apparent power} = \sqrt{(\text{true power})^2 + (\text{reactive power})^2}$$

$$\text{power factor} = \frac{Z}{R} = \frac{I_R}{I_S} = \cos \phi, \; \text{lagging}$$

$\phi = \text{inv} \cos (\text{power factor})$; ϕ is positive because i_S lags e

$$\sin \phi = \frac{Z_T}{X_L}, \; \tan \phi = \frac{R}{X_L}$$

$$B_L = \frac{1}{X_L}, \; G = \frac{1}{R}, \; Y_T = \frac{1}{Z_T} = \sqrt{G^2 + B_L{}^2} \text{ siemens}$$

☐ **C and R in Parallel**

$$Z_T = \frac{R \times X_C}{\sqrt{R^2 + X_C{}^2}} \; \Omega$$

$$I_S = \frac{E}{Z_T}, \; I_R = \frac{E}{R}, \; I_C = \frac{E}{X_C}$$

$$I_S = \sqrt{I_R{}^2 + I_C{}^2}$$

$$\text{true power} = \frac{E^2}{R} = I_R{}^2 \times R = E \times I_R = E \times I_S \times \text{p.f. watts}$$

$$\text{reactive power} = \frac{E^2}{X_C} = I_C{}^2 \times X_C = E \times I_C \text{ VAr}$$

$$\text{apparent power} = \frac{E^2}{Z_T} = I_S{}^2 \times Z_T = E \times I_S \text{ VA}$$

$$\text{apparent power} = \sqrt{(\text{true power})^2 + (\text{reactive power})^2}$$

$$\text{power factor} = \frac{Z_T}{R} = \frac{I_R}{I_S} = \cos \phi, \text{ leading}$$

$$\phi = \text{inv cos (power factor)}; \; \phi \text{ is negative because } i_S \text{ leads } e$$

$$\sin \phi = \frac{Z}{X_C}, \; \tan \phi = \frac{R}{X_C}$$

$$B_C = \frac{1}{X_C}, \; G = \frac{1}{R}, \; Y_T = \frac{1}{Z_T} = \sqrt{G^2 + B_C{}^2} \text{ siemens}$$

☐ **L and C in Parallel**

$$Z = \frac{X_L \times X_C}{X_L \sim X_C}, \; Y_T = \frac{1}{Z_T} = B_L \sim B_C$$

$$I_S = \frac{E}{Z}, \; I_L = \frac{E}{X_L}, \; I_C = \frac{E}{X_C}$$

$$I_S = \sqrt{I_L \sim I_C}$$

$$\text{true power} = 0 \text{ watts}$$

$$\text{reactive power} = I^2 Z = E I_S \text{ VAr}$$

power factor = 0 and is lagging if X_C is greater than X_L or leading if X_L is greater than X_C

ϕ equals $+90°$ if X_C is greater than X_L but is $-90°$ if X_L is greater than X_C

☐ **L, C and R in Parallel**

$$Z_T = \frac{R \times X}{\sqrt{R^2 + X^2}}, \text{ where } X = \frac{X_L \times X_C}{X_L \sim X_C}$$

$$Y_T = \frac{1}{Z_T} = \sqrt{G^2 + (B_L \sim B_C)^2}$$

$$I_S = \frac{E}{Z} = EY, \, I_R = \frac{E}{R} = EG, \, I_L = \frac{E}{X_L} = EB_L$$

$$I_C = \frac{E}{X_C} = EB_C$$

$$I_S = \sqrt{I_R^2 + (I_L \sim I_C)^2}$$

$$\text{true power} = \frac{E^2}{R} = E^2G = I_R^2 \times R = E \times I_R = E \times I_S \times \text{p.f. watts}$$

$$\text{total reactive power} = \frac{E^2}{X} \text{ VAr}$$

$$\text{apparent power} = \frac{E^2}{Z_T} = I_S^2 Z_T = EI_S \text{ VA}$$

$$\text{apparent power} = \sqrt{(\text{true power})^2 + (\text{total reactive power})^2}$$

$$\text{power factor} = \frac{Z}{R} = \frac{I_R}{I_S} = \cos\phi$$

The power factor is leading if X_L is greater than X_C but lagging if X_C is greater than X_L.

$$\phi = \text{inv} \cos (\text{power factor})$$

The sign of ϕ is positive if X_C is greater than X_L but negative if X_L is greater than X_C.

$$\sin\phi = \frac{Z_T}{X}, \, \tan\phi = \frac{R}{X}$$

☐ Maximum Power Transfer-Ac Case: Resistive Load
    P_ℓ has its maximum value when $R_\ell = \sqrt{R_S^2 + X_S^2} \, \Omega$

$$P_{\ell(max)} = \frac{E^2}{2(R_\ell + R_S)} \text{ watts}$$

☐ Power Transfer-Ac Case: Impedance Load
    P_ℓ has its maximum value when $R_\ell = R_S$ and $X_L = -X_S$. This means that the load and source impedances are conjugates.

$$P_{\ell(max)} = \frac{E^2}{4 \, R_S} = \frac{E^2}{4 \, R_\ell}$$

which occurs when the load current

$$I_\ell = \frac{E}{2 \, R_\ell} = \frac{E}{2 \, R_S}$$

11
Decibels and Nepers

In this chapter you will learn:

☐ How to convert power ratios into decibels that are based on common logarithms.

☐ How to reverse the process and convert decibels into power ratios.

☐ To calculate the overall decibel gain or loss for a number of cascaded stages.

☐ How to express an absolute power level in terms of decibels by using a reference level of 1 milliwatt.

☐ To convert voltage and current ratios into decibels (and vice-versa).

☐ How to measure a degree of attenuation in terms of nepers, which are based on natural logarithms.

In many instances, electronics is concerned with the transmission of alternating power from one position to another. The various lines and items of equipment that constitute the system introduce gains and losses of power.

POWER RATIOS

Consider a network connecting an alternating source to a load (Fig. 11-1). Let the input power be P_i and the output power be P_o. The ratio of output power to input power is then P_o/P_i. The network could introduce a loss, in which case P_o/P_i is less than unity, or it could introduce a gain (for example, an amplifier), in which case P_o/P_i is greater than unity.

If a number of such networks are connected in cascade (Fig. 11-2) and the individual power ratios are known, the overall power ratio P_o/P_i is

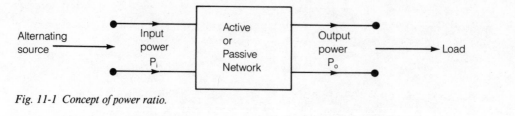

Fig. 11-1 Concept of power ratio.

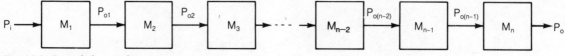

Fig. 11-2 Cascaded networks.

obtained by multiplying together the individual power ratios. This follows from the fact that

$$\frac{P_o}{P_i} = \frac{P_{o1}}{P_i} \times \frac{P_{o2}}{P_{o1}} \times \frac{P_{o3}}{P_{o2}} \times \cdots \times \frac{P_o}{P_{o(n-1)}}$$

or

$$M_T = M_1 \times M_2 \times M_3 \times \cdots \times M_n$$

where

M_1, etc., are the individual power ratios
M_T is the overall power ratio

Example 11-1

In Fig. 11-3, calculate the output power at Y if the input power at X is 1 mW.

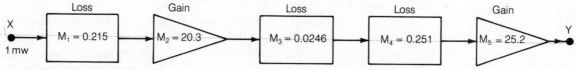

Fig. 11-3 Circuit for example 11-3.

Solution

overall power ratio, $M_T = 0.215 \times 20.3 \times 0.0246 \times 0.251 \times 25.2$
$= 0.679$ (a loss)

Therefore if 1 mW is applied to X, the output power at Y is 0.679 mW.

THE LOGARITHMIC UNIT

In a complex system containing a variety of circuits with each contributing a gain or loss, calculation of the overall power ratio can be extremely laborious. To simplify this calculation, the individual power ratios are expressed by a logarithmic unit, enabling addition to be employed in place of multiplication. The logarithmic unit employed is the "decibel" (abbreviated to dB), and the power gain or loss, N, or a network expressed in this unit is defined as

$$N = 10 \log_{10} \frac{P_o}{P_i}$$

where

$$P_o = \text{output power}$$
$$P_i = \text{input power}$$

If P_o/P_i is less than unity, then $10 \log_{10} P_o/P_i$ is negative. A negative sign therefore indicates a power loss, and a positive sign a gain.

It should be noted that because

$$10 \log_{10} \frac{P_o}{P_i} = -10 \log_{10} \frac{P_i}{P_o}$$

the numerical answer will be the same regardless of whether P_o/P_i or P_i/P_o is considered, but in order to obtain the correct sign, P_o/P_i must be used.

A larger logarithmic unit is the *Bel,* which is equivalent to 10 decibels and is named after Alexander Graham Bell. The decibel was originally related to acoustics and was regarded as the smallest change in sound intensity that could be detected by the human ear.

The following examples illustrate the use of the calculator in converting from power ratios to dB and vice-versa.

Example 11-2

(1) Convert power ratios of (a) 273 and (b) 0.0469 into decibels.
(2) Convert (a) +37 dB and (b) 3.0 dB into their corresponding ratios.

Solution

(1a) For the power ratio of 273, $N = 10 \log_{10} 273$ dB. On the calculator, enter "273", then press the "common" or "logarithm to the base 10" key (normally shown as "log"). The display will indicate 2.436, which must then be multiplied by 10 to give the required answer of 24.36 dB.

(1b) For a power ratio of 0.469, which represents a loss, $N = 10 \log_{10} 0.0469 = -13.3$ dB.

(2a) To convert +37 dB into its corresponding ratio, enter +37 and then divide by 10. The display indicates 3.7, after which the inverse logarithm key must be used. This key can be shown as "Inv. log," "$\log^{-1}$," "antilog" or "10^X", and when pressed, the required power ratio is displayed as 5012.

(2b) $\text{power ratio} = \text{inv log} \dfrac{(-3.0)}{10} = \text{inv log} \, (-0.3) = 0.5$

A loss of 3 dB is therefore equivalent to a power ratio of one-half.

Example 11-3

In Example 11-1, express each of the individual power ratios as well as the overall power ratio in terms of dB.

Solution

$$N_1 = 10 \log M_1 = 10 \log 0.215 = -6.68 \text{ dB}$$
$$N_2 = 10 \log M_2 = 10 \log 20.3 = +13.07 \text{ dB}$$
$$N_3 = 10 \log M_3 = 10 \log 0.246 = -16.09 \text{ dB}$$
$$N_4 = 10 \log M_4 = 10 \log 0.251 = -6.00 \text{ dB}$$
$$N_5 = 10 \log M_5 = 10 \log 25.2 = +14.01 \text{ dB}$$
$$N_T = 10 \log M_T = 10 \log 0.679 = -1.68 \text{ dB}$$

Then

$$\begin{aligned} N_T &= N_1 + N_2 + N_3 + N_4 + N_5 \\ &= -6.68 + 13.07 - 16.09 - 6.00 + 14.01 \\ &= -1.69 \text{ dB} \end{aligned}$$

The overall gain or loss in dB for a number of cascaded stages is therefore the algebraic sum of the decibel gains and losses in the individual stages.

EXPRESSION OF ABSOLUTE POWER LEVELS IN THE DECIBEL NOTATION

The decibel is fundamentally a unit of power ratio and not of absolute power, but if some standard reference level of power is assumed, then any absolute power can be expressed as so many dB above or below this reference standard. While various other standards might be encountered occasionally, the standard most commonly adopted is 1 mW (0.001 W). By using this standard, any power P can be expressed as

$$10 \log_{10} \frac{P}{1 \text{ mW}} \text{ dB}$$

referred to 1 mW. Therefore, 1 watt is

$$10 \log_{10} \frac{1000}{1} = 30 \text{ dB}$$

above the standard 1 mW, or "+30 dB with respect to 1 mW." Similarly, 5 μW is

$$10 \log_{10} \frac{5}{1000} = -23 \text{ dB}$$

with respect to 1 mW (23 dB below 1 mW). The expression "dB with respect to 1 mW" or "dB referred to 1 mW" is usually abbreviated to "dBm." Other abbreviations sometimes met are "dB wrt 1 mW." "dB ref 1 mW," and "VU" (volume or voice unit). It therefore appears that the dBm and the VU are the same; however, the use of the dBm is normally confined to single frequencies (tones) while VU is reserved for complex audio signals such as speech or music. However, both the dBm and the VU measure audio-power levels that are conveyed along a 600 standard program transmission line.

Example 11-4

(1) Express power levels of (a) 6.38 W and (b) 26.7 μW in dBm.
(2) What power levels are represented by (a) 617 dBm and (b) -8.6 VU?

Solution

(1a) Convert the power level to mW so that 6.38 W = 6.38 × 1000 mW = 6380 mW. Then

$$6.38 \text{ W} = 10 \log_{10}\left(\frac{6380 \text{ mW}}{1 \text{ mW}}\right) = 38.05 \text{ dBm}$$

(1b) $26.7 \mu\text{W} = 10 \log_{10}\left(\frac{26.7/1000 \text{ mW}}{1 \text{ mW}}\right)$

$$= 10 \log 0.0267$$
$$= -15.73 \text{ dBm}$$

(2a) power level $= \text{inv} \log\left(\frac{+17}{10}\right) = 50.12 \text{ mW}$

(2b) power level $= \text{inv} \log\left(\frac{-8.6}{10}\right)$

$$= 0.138 \text{ mW}$$
$$= 138 \mu\text{W}$$

CURRENT AND VOLTAGE RATIOS

When it is necessary to compare the powers delivered in equal input and output resistances, it is sufficient to measure their associated voltages or currents. The power ratio in decibels is then equal to twenty times the logarithm (to the base 10) of the current or voltage ratio.

Consider equal input and output resistances of R ohms, which carry currents of rms values I_i and I_o and develop voltages of rms values E_i and E_o respectively. The powers associated with the two resistances are

$$\text{input power, } P_i = E_i I_i = R I_i^2 = \frac{1}{R} \times E_i^2 \text{ watts}$$

output power, $P_o = E_o I_o = R I_o{}^2 = \dfrac{1}{R} \times E_o{}^2$ watts

The ratio between these two powers is therefore

$$\frac{P_o}{P_i} = \frac{E_o I_o}{E_i I_i} = \left(\frac{I_o}{I_i}\right)^2 = \left(\frac{E_o}{E_i}\right)^2$$

or, expressing this in decibels,

$$N = 10 \log_{10} \frac{P_o}{P_i} = 10 \log_{10} \left(\frac{I_o}{I_i}\right)^2 = 20 \log_{10} \frac{I_o}{I_i}$$

and

$$N = 10 \log_{10} \frac{P_o}{P_i} = 10 \log_{10} \left(\frac{E_o}{E_i}\right)^2 = 20 \log_{10} \frac{E_o}{E_i}$$

Therefore the power ratio in decibels is equal to $20 \log_{10}$ (current ratio) $= 20 \log_{10}$ (voltage ratio), provided that the input and output resistances are equal.

If the input and output powers are associated with input and output impedances of respectively $z_i = R_i + jX_i$ and $z_o = R_o + jX_o$, the formulas

$$N = 20 \log_{10} \frac{I_o}{I_i}$$

or

$$N = 20 \log_{10} \frac{E_o}{E_i}$$

still can be used for dB gain or loss provided $R_i = R_o$. If R_i and R_o are not equal, the formulas become

$$N = 10 \log_{10} \left(\frac{E_o{}^2/R_o}{E_i{}^2/R_i}\right) = 20 \log_{10} \frac{E_o}{E_i} - 10 \log_{10} \frac{R_o}{R_i}$$

$$= 20 \log_{10} \frac{E_o}{E_i} + 10 \log_{10} \frac{R_i}{R_o} \text{ dB}$$

and

$$N = 10 \log_{10} \left(\frac{I_o{}^2 R_o}{I_i{}^2 R_i}\right) = 20 \log_{10} \frac{I_o}{I_i} + 10 \log_{10} \frac{R_o}{R_i} \text{ dB}$$

Example 11-5

(1) Assuming that the input and output resistances are equal, convert (a) $+43.7$ dB and (b) -27.6 dB into their corresponding voltage (or current) ratios.

(2) Assuming equal input and output resistances, convert into dB (a) a voltage ratio of 56.3 and (b) a current ratio of 0.00353.

Solution

(1a) voltage (or current) ratio $= \text{inv log}\left(\dfrac{43.7}{20}\right)$

$= 153$

(2b) voltage (or current) ratio $= \text{inv log}\left(\dfrac{-27.6}{20}\right)$

$= 0.0417$

(2a) $N = 20 \log_{10} 56.3 = 35.01$ dB

(2b) $N = 20 \log_{10} 0.00353 = -49.04$ dB

Example 11-6

The input resistance of an amplifier is 1.3 kΩ while its output resistance is 17 kΩ. If the amplifier's voltage gain is 35, calculate its power gain in dB.

Solution

$$\text{power gain in dB} = 20 \log_{10} 35 - 10 \log_{10}\left(\dfrac{17}{1.3}\right)$$

$$= 30.88 - 8.59$$

$$= 22.3 \text{ dB}$$

EXPRESSION OF ATTENUATION IN NEPERS

The decibel is fundamentally a unit of power ratio, but it can be used to express current ratios when the resistive components of the impedances through which the currents flow are equal, and voltage ratios when the conductive components of the corresponding admittances are equal. The neper is fundamentally a unit of current ratio, but it can be used to express power ratios when the resistive components of the impedances are equal (Fig. 11-4).

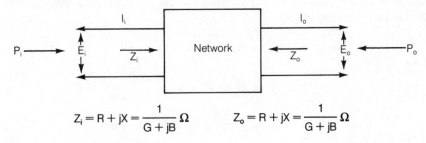

$$Z_i = R + jX = \frac{1}{G + jB}\ \Omega \qquad Z_o = R + jX = \frac{1}{G + jB}\ \Omega$$

Fig. 11-4 Decibels and nepers.

The loss of power in an electrical network is known as *attenuation*. Attenuation can be measured using either the decibel or the neper notation. If the power entering a network is P_i and the power leaving is P_o, then the attenuation in decibels is defined as

$$\text{attenuation in decibels} = 10 \log_{10} \frac{P_o}{P_i}$$

If the current entering a network is I_i and the current is I_o, then the attenuation in nepers is defined as

$$\text{attenuation in nepers} = \ln \frac{I_o}{I_i}$$

where ln is the natural or Napierian logarithm, which is based on the exponential value, $e = 2.7183 \ldots$

Because of its derivation from the exponential value, e, the neper is the most convenient unit for expressing attenuation in theoretical work. The decibel, on the other hand, being defined in terms of logarithms to the base 10, is a more convenient unit in practical calculations using the decimal system. The conditions under which the two units can be used are summarized in the following equations:

$$\text{attenuation in dB} = 10 \log_{10} \frac{P_o}{P_i}$$

$$= 20 \log_{10} \frac{I_o}{I_i}, \text{ provided that } R_i = R_o$$

$$= 20 \log_{10} \frac{E_o}{E_i}, \text{ provided that } G_i = G_o$$

$$\text{attenuation in nepers} = \ln \frac{I_o}{I_i} = \ln \frac{E_o}{E_i}, \text{ provided that } Z_i = Z_o$$

$$= \frac{1}{2} \ln \frac{P_o}{P_i} \text{ provided that } R_i = R_o$$

If the resistive components of the impedances at the input and output of the network are equal, then the attenuation can be converted readily from one notation to the other.

$$\text{attenuation in dB} = 20 \log_{10} \frac{I_o}{I_i}$$

$$= 20 \ln \frac{I_o}{I_i} \times \log_{10} e$$

$$= 8.686 \times \ln \frac{I_o}{I_i}$$

$$= 8.686 \times (\text{attenuation in nepers})$$

Therefore attenuation in dB = $8.686 \times$ (attenuation in nepers) or attenuation in nepers = $0.1151 \times$ (attenuation in dB) provided that $R_i = R_o$.

Example 11-7

The input current of an electrical network is 27 mA and the output current is 75 μA. Assuming that the input and output resistances are equal, calculate the attenuation of the network in (a) decibels and (b) nepers.

Solution

(a)
$$\text{current ratio} = \frac{I_o}{I_i} = \frac{75\ \mu A}{27\ mA} = 2.77 \times 10^{-3}$$

$$\text{attenuation in dB} = 20 \log_{10}(2.77 \times 10^{-3})$$
$$= -51.1\ dB$$

(b)
$$\text{attenuation in nepers} = \ln(2.77 \times 10^{-3})$$
$$= -5.889\ \text{nepers}$$

Check:

$$\frac{\text{attenuation in dB}}{\text{attenuation in nepers}} = \frac{51.1}{5.889} = 8.68, \text{ rounded off}$$

CHAPTER SUMMARY

☐ Power Ratio

$$N = 10 \log_{10} \frac{P_o}{P_i}\ dB$$

☐ Cascaded Stages

$$N_T = N_1 + N_2 + N_3 \cdots dB$$

☐ Absolute Power

$$N = 10 \log_{10} \frac{P}{1\ mW}\ dBm$$

☐ Voltage Ratio

$$N = 20 \log_{10} \frac{E_o}{E_i}\ dB, \text{ provided } R_i = R_o$$

☐ Current Ratio

$$N = 20 \log_{10} \frac{I_o}{I_i}\ dB, \text{ provided } R_i = R_o$$

☐ Nepers

$$N = \ln \frac{I_o}{I_i}$$

$$= \ln \frac{E_o}{E_i} \text{ provided } Z_i = Z_o$$

$$= \frac{1}{2} \ln \frac{P_o}{P_i} \text{ provided } R_i = R_o$$

attenuation in decibels = 8.686 × (attenuation in nepers) provided
$$R_i = R_o$$
attenuation in nepers = 0.1151 × (attenuation in decibels) provided
$$R_i = R_o$$

12
Resonance

IN OUR DISCUSSION on the subject of resonance, you will learn:

☐ The definition of resonance as it applies to any ac circuit with only one source.

☐ About the properties of series resonant LCR circuits.

☐ About the merit factor in Q, the series LCR circuit and its interpretations as the voltage magnification factor, the degree of selectivity and the reciprocal of the inductor's power factor.

☐ The meaning of the term "bandwidth" and its relation to the Q factor.

☐ About the properties of parallel resonant LCR circuits.

☐ How to interpret the Q factor for a parallel LCR circuit in its relation to current magnification, impedance magnification, and the degree of selectivity.

☐ About the properties of a parallel resonant tank circuit.

There is more than one way to define the condition of resonance. This book uses the following definition: any two-terminal network (this means that there is only one supply) containing resistance and reactance is said to be in resonance when the supply voltage and the current drawn from the supply are in phase (the phase angle of the circuit is zero).

It follows from this definition that a resonant circuit has a power factor of unity.

RESONANCE IN A SERIES LCR CIRCUIT

From the definition of resonance, the phase angle, ϕ, is zero. Referring to Fig. 12-1, this requires that $X_L = X_C$, and therefore $V_L = V_C$; the phasor

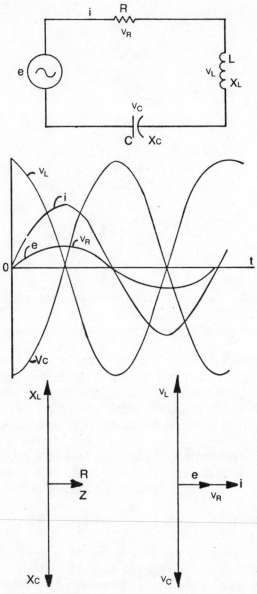

Fig. 12-1 Series-resonant LCR circuit.

sum of v_L and v_C is zero, so that $e = v_R$ and the circuit is purely resistive. The true power is $EI = I^2R = E^2/R$ watts.

In terms of magnitudes, $E = V_R$ and $Z = R$. This means that the total impedance of the circuit has a minimum value at resonance and, therefore, the corresponding current, $I = E/R$, is at its maximum. For these reasons, the series LCR circuit at resonance is sometimes referred to as an *acceptor circuit*. With the given values of L and C, resonance occurs at a particular frequency:

$$f_r = \frac{1}{2\,\pi\,\sqrt{LC}} = \frac{0.159}{\sqrt{LC}}$$

where

f_r is measured in Hz
L is in henrys
C is in farads.

Then

$$L = \frac{1}{4\,\pi^2 f_r^2 C} = \frac{0.0253}{f_r^2 C}\ H$$

and

$$C = \frac{1}{4\,\pi^2 f_r^2 L} = \frac{0.0253}{f_r^2 L}\ F$$

Note that the value of the resonant frequency, f_r, is related to the product of L and C but is independent of R. Tuning a series LCR circuit means adjusting the value of L or C until the resonant frequency is equal to the desired signal frequency. The behavior of the series LCR circuit as the frequency is varied is illustrated by means of response curves. These are graphs of particular circuit variables that are plotted against frequency. The most common response curves (Fig. 12-2A and 12-2B) are those for the total impedance, Z, and the circuit current, I.

At frequencies below the value of f_r, X_C is greater than X_L, i leads e, and the circuit behaves capacitively. At frequencies above f_r, X_L is greater than X_C, and i lags e; the circuit then behaves inductively.

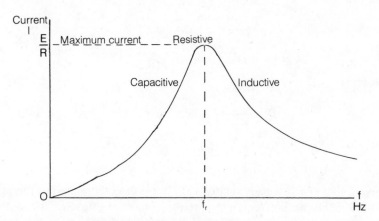

Fig. 12-2. A Current response curve of the series LCR circuit.

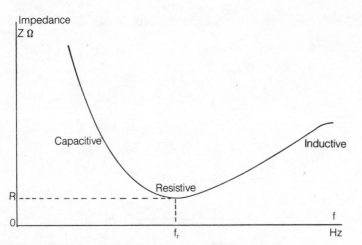

Fig. 12-2 B Impedance response curve of the series LCR circuit.

The quantity "*Q*" can be regarded as a merit factor related to the inductor in a tuned circuit. It can be defined as the ratio of the inductor's reactance to its resistance so that $Q = X_L/R$ and is therefore a number without any units. The Q factor associated with a series tuned circuit has certain interpretations, of which the most important follow.

Voltage Magnification Factor

At resonance, the circuit current has its maximum value, and equal (but 180° out-of-phase) voltages are developed across the inductor and the capacitor. These voltages each can be many times greater than the applied voltage (refer to Example 12-1). The number of times greater is called the *magnification factor*, and is equal to Q.

Therefore at resonance

$$Q = \frac{V_L}{E} = \frac{V_C}{E}$$

$$= \frac{IX_L}{IR} = \frac{X_L}{R} = \frac{2\pi f_r L}{R}$$

Because

$$f_r = \frac{1}{2\pi\sqrt{LC}}$$

$$Q = \frac{2\pi L}{2\pi\sqrt{LC} \times R} = \frac{1}{R} \times \sqrt{\frac{L}{C}}$$

Note that Q is inversely related to R but is directly related to the square root of L/C. For audio frequency circuits, Q is of the order of 10, but with radio frequency circuits, the value of Q can exceed 100.

Reciprocal of the Inductor's Power Factor

$$\text{inductor's power factor} = \frac{R}{Z}$$

$$= \frac{R}{\sqrt{R^2 + X_L{}^2}}$$

$$= \frac{R}{X_L \times \sqrt{1 + \dfrac{R^2}{X_L{}^2}}}$$

$$= \frac{1}{Q \times \sqrt{1 + \dfrac{1}{Q^2}}}$$

which, to within 1%, is equal to $1/Q$, provided Q is greater than 10. Therefore, the values of Q and the power factor for radio frequency inductors are reciprocals.

Selectivity

The selectivity of a series-tuned circuit is defined as its ability to distinguish between the signal frequency to which it is resonant and other signals on nearby frequencies. It therefore follows that the greater the selectivity, the greater is the freedom from adjacent channel interference. The degree of selectivity is related to the sharpness of the current response curve (the sharper the curve, the greater the selectivity) and can be measured by the frequency separation between two specific points on the curve (Fig. 12-3). The points usually chosen are those for which the true power in the circuit is half of the maximum true power, which occurs when the circuit is resonant. These positions on the response curve are often referred to as the *3 decibel (dB) points* (a loss of 3 dB is equivalent to a power ratio of 1/2; see Chapter 11) and their frequency separation is called the *bandwidth* (or bandpass) value of the tuned circuit. At the 3-dB points, the rms circuit current is $1/\sqrt{2}$,

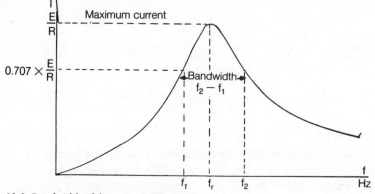

Fig. 12-3 Bandwidth of the series LCR circuit.

or 0.707 times the rms value of the current at resonance (do not confuse this result with the relationship between the rms and the peak values of a sine-wave alternating current). In addition, at the 3-dB points, the net (total) circuit reactance is equal to the circuit resistance so that the phase angle is 45° and the power factor is 0.707.

It can be shown that the bandwidth is

$$f_2 - f_1 = \frac{R}{2 \pi L} \text{ Hz}$$

and therefore

$$\frac{\text{bandwidth}}{\text{resonant frequency}} = \frac{R}{2 \pi f_r L} = \frac{R}{X_L} = \frac{1}{Q}$$

or

$$\text{Bandwidth} = \frac{\text{resonant frequency}}{Q} \text{ and } Q = \frac{\text{resonant frequency}}{\text{bandwidth}}$$

Therefore Q is a direct measure of the degree of selectivity. The higher the value of Q, the sharper the current response curve, the greater the degree of selectivity, and the narrower the bandwidth.

Energy Relationships

Q is equal to 2π times the ratio of the maximum energy stored during the cycle to the mean energy dissipated over the period.

The maximum energy stored during the cycle in the form of the magnetic field surrounding the inductor, is $\frac{1}{2} L(I_{peak})^2$, where I_{peak} is the peak value of the current at resonance. The maximum energy is therefore $\frac{1}{2} L(\sqrt{2}I_{rms})^2 = L(I_{rms})^2$ joules. The mean energy dissipated over the period T is equal to power $\times$ time, or

$$\text{mean energy} = (I_{rms})^2 R T$$

$$= (I_{rms})^2 R \times \frac{1}{f_r} \text{ joules}$$

Then

$$2 \pi \times \frac{\text{maximum energy stored during the cycle}}{\text{mean energy dissipated over the period}}$$

$$= 2 \pi \times \frac{L(I_{rms})^2}{(I_{rms})^2 \times R/f_r} = 2 \pi \times \frac{f_r L}{R} = Q$$

Note that if two identical series LCR circuits are joined end-to-end with zero mutual coupling between the coils, the total impedance of the series combination at resonance is doubled ($Z = 2R$), but the Q and the resonant frequency remains the same. However, if a single additional damping resistor. R_d is connected to the series LCR circuit, the new impedance at resonance is greater and equal to $R + R_d$ so that the new Q is given by

$$Q_{new} = \frac{1}{R + R_d} \times \sqrt{\frac{L}{C}} = \frac{R}{R + R_d} \times Q_{old}$$

and is reduced. New bandwidth = $(R + R_d/R) \times$ old bandwidth and is increased, but the resonant frequency is unchanged.

Example 12-1

In Fig. 12-1, R = 8 Ω, L = 150 μH, C = 250 pF, and E = 1.4 V. Calculate the resonant frequency and determine the resonant values of Z, I, V_R, V_L, V_C, and the true power. What are the values of Q and the circuit's bandwidth? If an additional 12 Ω resistor is inserted in series, calculate the new values of the resonant frequency, Q and the bandwidth.

Solution

The resonant frequency

$$f_r = \frac{1}{2 \pi \sqrt{LC}}$$

$$= \frac{0.159}{\sqrt{150 \times 10^{-6} \times 250 \times 10^{-12}}} \text{ Hz}$$

$$= 822 \text{ kHz}$$

At the resonant frequency,

$$X_L = X_C = 2 \times \pi \times 822 \times 10^3 \times 150 \times 10^{-6}$$
$$= 775 \text{ }\Omega$$

At resonance,

$$Z = R = 8 \text{ }\Omega$$

$$I = \frac{E}{R} = \frac{1.4 \text{ V}}{8 \text{ }\Omega} = 175 \text{ mA}$$

$$V_R = E = 1.4 \text{ V}$$
$$V_L = IX_L = 175 \text{ mA} \times 775 \text{ }\Omega = 136 \text{ V}$$
$$V_C = V_L = 136 \text{ V}$$
$$\text{true power} = I^2R = (175 \text{ mA})^2 \times 8 \text{ }\Omega = 0.245 \text{ W}$$

The power factor is unity, and the phase angle between e and i is zero.

$$Q = \frac{1}{R} \times \sqrt{\frac{L}{C}} = \frac{1}{8} \times \sqrt{\frac{150 \times 10^{-6}}{225 \times 10^{-12}}} = 97$$

$$\text{Check: } Q = \frac{X_L}{R} = \frac{X_C}{R} = \frac{775 \text{ }\Omega}{8 \text{ }\Omega} = 97.$$

$$\text{and } Q = \frac{V_L}{E} = \frac{V_C}{E} = \frac{136 \text{ V}}{1.4 \text{ V}} = 97.$$

$$\text{bandwidth} = \frac{R}{2 \pi L} = \frac{8}{2 \times \pi \times 150 \times 10^{-6}} \text{ Hz} = 8.5 \text{ kHz}$$

$$\text{Check: bandwidth} = \frac{f_r}{Q} = \frac{822 \text{ kHz}}{97} = 8.5 \text{ kHz}$$

When the additional 12 Ω resistor is inserted in series, the resonant frequency is unchanged. The new Q is

$$97 \times \frac{8 \text{ Ω}}{8 \text{ Ω} + 12 \text{ Ω}} = 39$$

and the new bandwidth is

$$\frac{822}{39} = 21.2 \text{ kHz}$$

RESONANCE IN A PARALLEL LCR CIRCUIT

The phase relationships between e, i_R, i_L, and i_c have already been described. At resonance, the phase angle, ϕ, between e and i_s, is zero; this requires that $I_L = I_C$, and $I_S = I_R$ (Fig. 12-4). The total impedance of the circuit, Z_T, is resistive and also has a maximum value that is equal to R (note that the circuit impedance can never exceed R). The supply current, I_S, is a minimum and equal to E/R. The resonant frequency is

$$f_r = \frac{1}{2 \pi \sqrt{LC}} = \frac{0.159}{\sqrt{LC}} \text{ Hz}$$

which is the same expression as for series resonance.
Then

$$L = \frac{1}{4 \pi^2 f_r^2 C} = \frac{0.0253}{f_r^2 C} \text{ H}$$

and

$$C = \frac{1}{4 \pi^2 f_r^2 L} = \frac{0.0253}{f_r^2 L} \text{ F}$$

The response curves for the parallel LCR circuit appear in Figs. 12-5A and 12-5B.

At frequencies below the resonant frequency f_L, X_C is greater than X_L, I_C is less than I_L, and the circuit behaves inductively. At frequencies above the resonance frequency, X_L is greater than X_C, I_L is less than I_C, and the circuit behaves capacitively. These results are the reverse of those for the series LCR circuit, which behaves capacitively for frequencies below f_r and inductively for frequencies above f_r.

The Q factor is interpreted in the following sections:

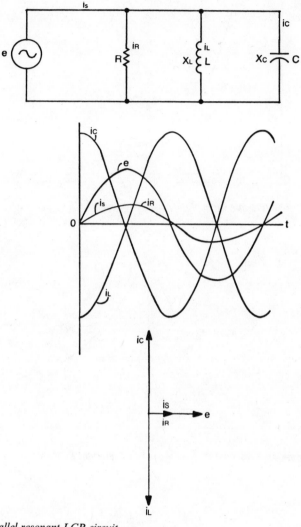

Fig. 12-4 *Parallel-resonant LCR circuit.*

Current Magnification

At resonance, the currents I_L and I_C are equal in magnitude, and each is Q times the supply current, I_S.
Then

$$I_L = \frac{E}{X_L}, \ I_C = \frac{E}{X_C}, \ I_S = I_R = \frac{E}{R}$$

$$Q = \frac{I_L}{I_S} = \frac{E/X_L}{E/R} = \frac{R}{X_L} = \frac{R}{X_C} \left(\text{not} \ \frac{X_L}{R} \text{ or } \frac{X_C}{R} \right)$$

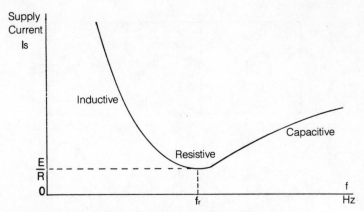

Fig. 12-5 A *Current response curve of the parallel-resonant LCR circuit.*

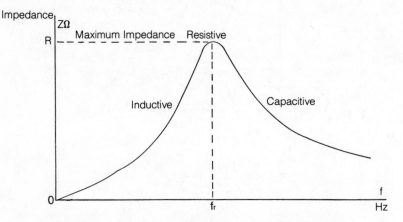

Fig. 12-5 B *Impedance response curve of the parallel-resonant LCR circuit.*

Because

$$Q = \frac{R}{X_L} = \frac{R}{2 \pi f_r L}$$

and

$$f_r = \frac{1}{2 \pi \sqrt{LC}}$$

$$Q = \frac{R}{2 \pi L / 2 \pi \sqrt{LC}} = R \times \sqrt{\frac{C}{L}}$$

Impedance Magnification

Because $R = QX_L = QX_C$, the impedance at resonance is Q times the reactance of either the inductor or the capacitor.

Selectivity

As in the case in the series LCR circuit, Q determines the sharpness of the response curves and is therefore a direct measure of selectivity.

$$Q = \frac{\text{resonant frequency}}{\text{bandwidth}}$$

and the bandwidth can be defined for the impedance response curve as the frequency separation between the points at which the total impedance of the current is $0.707 \times$ the maximum impedance, R. From this definition:

$$\text{bandwidth} = \frac{1}{2\,\pi\,CR}\ \text{Hz}$$

where C is measured in farads and R in ohms. The concept of the half-power points cannot be used in the case of the parallel LCR circuit since the true power is always E^2/R and is independent of frequency.

Energy Relationships

As in the case in the series LCR circuit, Q can be defined as

$$Q = 2\,\pi \times \frac{\text{maximum energy stored during the cycle}}{\text{mean energy dissipated over the period}}$$

The maximum energy stored in the capacitor is

$$\frac{1}{2}\,C(E_{peak})^2 = \frac{1}{2}\,C(\sqrt{2}E_{rms})^2 = C(E_{rms})^2$$

The mean energy dissipated during the period is

$$\frac{(E_{rms})^2}{R} \times T$$

where T is the period, which is equal to $\dfrac{1}{f_r}$

Then

$$Q = 2\,\pi \times \frac{C(E_{rms})^2}{T(E_{rms})^2/R} = R \times 2\,\pi\,f_r C = \frac{R}{X_C}$$

Note that if two identical parallel LCR circuits are shunted across each other with zero mutual coupling between the coils, the effective resistance is halved. The impedance at resonance is therefore halved, but the Q, the resonant frequency, and the bandwidth remain the same.

If a single damping resistor, R_d, is connected across the parallel LCR circuit, the new impedance at resonance is

$$\frac{R_d \times R}{R + R_d}$$

and is reduced.
Then

$$Q_{new} = \frac{R_d \times R}{R + R_d} \times \sqrt{\frac{C}{L}} = \frac{R_d}{R + R_d} \times Q_{old}$$

and the value of Q therefore has been decreased.

$$new\ bandwidth = \frac{R + R_d}{R_d} \times old\ bandwidth$$

and is consequently increased, but the resonant frequency remains the same.

Example 12-2

In Fig 12-4, $R = 8.2\ k\Omega$, $L = 3.3\ \mu H$, $C = 6.5\ pF$, and $E = 15\ V$. Calculate the resonant frequency and determine the resonant values of I_S, I_R, I_C, and the true power. What are the values of Q and the bandwidth of the circuit. If an additional 10 kΩ resistor is added in parallel, what are the new values of the resonant frequency, Q, and bandwidth?

Solution

resonant frequency, $f_r = \dfrac{1}{2\ \pi\ \sqrt{LC}}$

$$= \frac{1}{2 \times \pi \times \sqrt{3.3 \times 10^{-6} \times 6.5 \times 10^{-12}}} H = 34.4\ MHz$$

At resonance, the inductive reactance, $X_L\ (= X_C)$ is

$$2 \times \pi \times 34.4 \times 10^6 \times 3.3 \times 10^{-6} = 712.5\ \Omega$$

$$I_R = I_S = \frac{E}{R} = \frac{15\ V}{8.2\ k\Omega} = 1.83\ mA$$

$$I_L = I_C = \frac{15\ V}{712.5\ \Omega} = 21.05\ mA$$

$$true\ power = \frac{E^2}{R} = \frac{(15\ V)^2}{8.2\ k\Omega} = 27.4\ mW$$

$$Q = R \times \sqrt{\frac{C}{L}} = 8.2 \times 10^3 \times \sqrt{\frac{6.5 \times 10^{-12}}{3.3 \times 10^{-6}}} = 11.5$$

Check:

$$Q = \frac{I_C}{I_S} = \frac{R}{X_C} = \frac{8.2 \times 10^3}{712.5} = 11.5$$

$$bandwidth = \frac{1}{2\ \pi\ CR} = \frac{1}{2 \times \pi \times 6.5 \times 10^{-12} \times 8.2 \times 10^3}\ Hz$$

$$= 2.99 MHz$$

Check:

$$\text{bandwidth} = \frac{f_r}{Q} = \frac{34.4 \text{ MHz}}{11.5} = 2.99 \text{ MHz}$$

When the additional 10 kΩ resistor is added in parallel, the new total resistance is

$$\frac{10 \times 8.2}{10 + 8.2} = 4.5 \text{ k}\Omega$$

The resonant frequency remains unchanged but the new Q is

$$11.5 \times \frac{4.5}{8.2} = 6.3$$

and the new bandwidth is

$$2.99 \times \frac{8.2}{4.5} = 5.45 \text{ MHz}$$

PARALLEL RESONANT TANK CIRCUIT

This circuit consists of a practical coil with resistance in parallel with a capacitor whose losses are assumed to be negligible (Fig. 12-6). Such a circuit is commonly used as the collector load of certain rf amplifiers.

Because the inductor branch contains both inductive reactance and resistance, current i_L lags e by a phase angle, ϕ. If the frequency is raised, the impedance of the inductor branch will increase, i_L will decrease, and the angle, ϕ, will approach 90°.

Capacitor current i_C leads e by 90°, and supply current i_S is the phasor sum of i_L and i_C. Figure 12-7 illustrates changes in the circuit's behavior as the frequency is varied. At all frequencies below the resonant frequency f_r, i_S

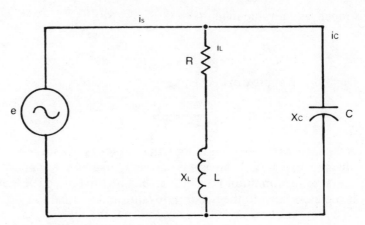

Fig. 12-6 Parallel-resonant tank circuit.

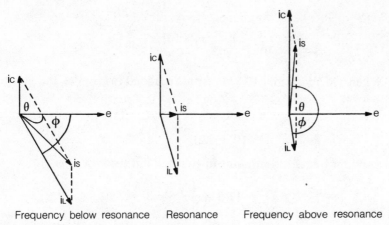

Frequency below resonance   Resonance   Frequency above resonance

Fig. 12-7 Phasor diagrams of the parallel tank circuit.

lags e by the total phase angle, θ, and the circuit behaves inductively. At all frequencies above f_r, i_S leads e and the circuit is capacitive. These results are similar to those obtained for the parallel LCR circuit discussed in Chapter 11 but are opposite to those for the series LCR circuit discussed earlier in that chapter.

At resonance, e and i_S are in phase so that $\theta = 0°$. For resonant tank circuits possessing a high Q, I_L can be considered as equal to I_C and X_L as equal to X_C. However, the exact expression for the resonant frequency, f_r, is

$$f_r = \frac{1}{2\pi} \times \sqrt{\frac{1}{LC} - \frac{R^2}{L^2}} \text{ Hz}$$

For rf tank circuits possessing an appreciable Q (greater than 10), the

$$\frac{R^2}{L^2}$$

term is much less in value than the

$$\frac{1}{LC}$$

term. Therefore, f_r approximately equals

$$\frac{1}{2\pi \times \sqrt{LC}}$$

which is the same as the expression for f_r in Chapter 11. As suggested by the phasor diagram in Fig. 12-7, the supply current I_S is small at resonance and is nearly equal to its minimum possible value. The total impedance at resonance is therefore close to the maximum value and is equal to

$$\frac{L}{CR} \, \Omega$$

From the definition of resonance, this quantity is resistive and is called the *dynamic resistance;* the term *dynamic* means that

$$\frac{L}{CR}\ \Omega$$

appears only under operating conditions. The parallel tank is sometimes referred to as a *rejector circuit* because it presents virtually maximum impedance at resonance. The principal response curves of the tank circuit appear in Figs. 12-8A and 12-8B.

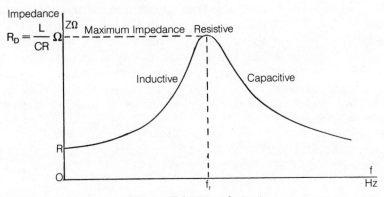

Fig. 12-8 A Response curve of the parallel LCR tank circuit.

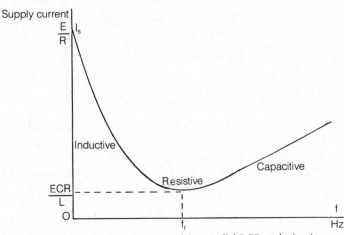

Fig. 12-8 B Supply current response curve of the parallel LCR tank circuit.

The Q factor can be interpreted several different ways, which are discussed below.

Impedance magnification   The impedance at resonance is Q times the reactance of the capacitor branch. Therefore

$$Q = \frac{L/(CR)}{1/(2\ \pi\ f_r C)} = \frac{2\ \pi\ f_r L}{R} = \frac{X_L}{R}$$

which is the same expression obtained for the series LCR resonant circuit.

Because X_L and X_C are nearly equal at resonance, $Q = X_C/R$, and for the impedance at resonance, $Z = X_C Q \approx Q^2R$ (the symbol "$\approx$" means "is approximately equal to").

Current magnification   Because i_L and i_C are nearly equal in magnitude and approximately $180°$ out of phase, there appears to be a large circulating, or *flywheel*, current, existing between L and C. The supply, or "make-up," current, I_S, is relatively small because it merely accounts for the fact that i_L and i_C are not exactly equal in magnitude and $180°$ out of phase. The ideal case of $R = 0$ was covered earlier, which showed that when $X_L = X_C$, the total impedance and Q were both infinite, and the supply current I_S was zero.

In the practical case, the impedance at resonance is Q times the reactance of the capacitor branch and is approximately equal to Q times the impedance of the inductor branch. Therefore the circulating current is Q times the supply current.

The true power in the circuit at resonance is $I_S^2 \times$ the dynamic resistance $= I_L^2R$, or

$$\frac{\text{circulating current, } I_L}{\text{make-up current, } I_S} = \sqrt{\frac{L/(CR)}{R}} = \frac{1}{R} \times \sqrt{\frac{L}{C}} = Q$$

Selectivity   Q is again a measure of selectivity and equals f_r/bandwidth. With the impedance response curve, the bandwidth can be defined as the frequency separation between the points at which the total circuit impedance is 0.707 times the impedance at resonance.

Energy relationships   Here again

$$Q = 2\pi \times \frac{\text{maximum energy stored during the cycle}}{\text{mean energy dissipated over the period}}$$

Note that if two identical tank circuits are paralleled with zero mutual coupling between the coils, the impedance at resonance is halved, but the Q and the resonant frequency remain unchanged. If a single damping resistor, R_d, is connected across the tank circuit, the new impedance at resonance is

$$Z_{new} = \frac{R_d \times L/(CR)}{R_d + L/(CR)}$$

and

$$Q_{new} = \frac{R_d}{R_d + L/(CR)} \times Q_{old}$$

so that both quantities are reduced. The bandwidth is therefore greater, but the resonant frequency remains the same (assuming that Q is greater than 10).

Example 12-3

In Fig. 12-6, R = 18 Ω, L = 25 μH, C = 15 pF, and E = 100 mV. Determine the resonant frequency and calculate the resonant values of I_S, I_L, and I_C. What are the values of Q and the tank circuit's bandwidth?

Solution

The exact formula for the resonant frequency is

$$f_r = \frac{1}{2\pi} \times \sqrt{\frac{1}{LC} - \frac{R^2}{L^2}}$$

However

$$\frac{1}{LC} = \frac{1}{25 \times 10^{-6} \times 15 \times 10^{-12}} = 2.667 \times 10^{15}$$

and

$$\frac{R^2}{L^2} = \frac{324}{(25 \times 10^{-6})^2} = 5.18 \times 10^{11}$$

Therefore R^2/L^2 is negligible when compared with $1/LC$.
The formula then becomes

$$f_r = \frac{1}{2\pi \times \sqrt{LC}} = \frac{1}{2 \times \pi \times \sqrt{25 \times 10^{-6} \times 15 \times 10^{-12}}}$$
$$= 8.22 \text{ MHz}$$

At the resonant frequency,

$$X_C = \frac{1}{2\pi f_r C} = \frac{1}{2 \times \pi \times 8.22 \times 10^6 \times 15 \times 10^{-12}} = 1291 \ \Omega$$

The impedance of the inductor branch is $\sqrt{R^2 + X_L^2} = \sqrt{18^2 + 1291^2} \approx 1291 \ \Omega$.

$$\text{circulating current, } I_L = I_C = \frac{100 \text{ mV}}{1291 \ \Omega} = 0.07746 \text{ mA}$$
$$= 77.46 \ \mu A$$

$$\text{dynamic resistance, } R_D = \frac{L}{CR} = \frac{25 \times 10^{-6}}{15 \times 10^{-12} \times 18} \ \Omega$$
$$= 92.6 \text{ k}\Omega$$

$$\text{supply or make-up current, } I_S = \frac{100 \text{ mV}}{92.6 \text{ k}\Omega} = 1.08 \ \mu A$$

$$Q = \frac{X_L}{R} = \frac{1291}{18} = 71.7$$

$$\text{Check: } Q = \frac{I_C}{I_S} = \frac{77.46}{1.08} = 71.7$$

$$\text{bandwidth} = \frac{f_r}{Q} = \frac{8.22}{71.7} \text{ MHz} = 115 \text{ kHz}$$

CHAPTER SUMMARY

☐ Series Resonant LCR Circuit
$\phi = 0°$. Power factor is unity.

$$\text{resonant frequency, } f_r = \frac{1}{2\,\pi \times \sqrt{LC}} \text{ Hz}$$

$$L = \frac{0.0253}{f_r^2 C} \text{ H, } C = \frac{0.0253}{f_r^2 L} \text{ F}$$

$$Z = R(\text{minimum value}), I = \frac{E}{R} \text{ (maximum value)}$$

$$V_R = E, V_L = V_C = Q \times E$$

$$Q = \frac{V_L}{E} = \frac{V_C}{E} = \frac{X_L}{R} = \frac{X_C}{R} = \frac{1}{R} \times \sqrt{\frac{L}{C}}$$

$$= \frac{1}{\text{inductor's power factor}}$$

$$\text{bandwidth} = \frac{R}{2\,\pi\,L} \text{ Hz} = \frac{f_r}{Q}$$

Current at 3 dB points = 0.707 × resonant current.
Maximum power at resonance = $I^2 R$ watts.
☐ Parallel Resonant LCR Circuit
Circuit phase angle is zero. Power factor is unity.

$$\text{resonant frequency, } f_r = \frac{1}{2\,\pi\,\sqrt{LC}} \text{ Hz}$$

$$Z_T = R(\text{maximum value}), I_S = \frac{E}{R} \text{ (minimum value)}.$$

$$I_L = I_C = Q \times I_S.$$

$$Q = \frac{I_L}{I_S} = \frac{I_C}{I_S} = \frac{R}{X_L} = \frac{R}{X_C} = R \times \sqrt{\frac{C}{L}}$$

$$\text{bandwidth} = \frac{f_r}{Q} = \frac{1}{2\,\pi\,RC} \text{ Hz}$$

Impedance at bandwidth points = 0.707 × R.
☐ Parallel Resonant Tank Circuit

Circuit phase angle is zero. Power factor is unity.

$$\text{resonant frequency, } f_r = \frac{1}{2\pi} \sqrt{\frac{1}{LC} - \frac{R^2}{L^2}} \approx \frac{1}{2\pi\sqrt{LC}} \text{ Hz}$$

provided Q is greater than 10.

$$Z_T = R_D = \frac{L}{CR} = Q \times X_C = Q \times X_L \approx Q^2 R \ \Omega.$$

$$I_S = \frac{ECR}{L}$$

$$Q = \frac{X_L}{R} = \frac{R_D}{X_L} = \frac{R_D}{X_C} = \frac{I_L}{I_S} = \frac{I_C}{I_S} = \frac{1}{R} \times \sqrt{\frac{L}{C}}$$

$$\text{bandwidth} = \frac{f_r}{Q}$$

13
Complex Algebra

IN STUDYING THE subject of complex algebra you will learn:

☐ About the importance of the j operator in the analysis of ac circuits.
☐ How the use of the j operator allows you to distinguish between the effects of resistance and reactance and to distinguish between real, imaginary, and complex quantities.
☐ About the derivation of the result that $j^2 = -1$.
☐ How to convert between the phasor rectangular and polar notations.
☐ To use the rules of complex algebra in order to add, subtract, multiply, and divide two or more phasors.
☐ To obtain the square, square root, and reciprocal of a phasor quantity.
☐ To use the j operator to calculate the total impedance of a series-parallel network.
☐ To use the j operator to calculate the total impedance of a number of parallel branches.

In previous chapters, the alternating voltage or current has been represented by a sine waveform, a phasor, or a trigonometric expression. Although these representations are adequate for simple series and parallel arrangements, they are too cumbersome for the analysis of more complicated circuits that require the use of the network theorems. What is needed is a form of algebra that can be applied directly to the solution of ac circuits. Such an algebra must be capable of taking into account the circuits' phase relationships by distinguishing between the resistive and the reactive elements. This is achieved in complex algebra by the introduction of the *j operator*.

INTRODUCTION TO COMPLEX ALGEBRA
AND TO THE j OPERATOR

When multiplied by the operator j, a phasor is rotated through 90° or $\pi/2$ radians in the positive or counterclockwise direction, but the magnitude of the phasor is unchanged.

Referring to Fig. 13-1, OP represents a phasor in the horizontal reference position. From the definition of the j operator,

$$\text{phasor OP}_1 = j \times \text{phasor OP}$$
$$\text{phasor OP}_2 = j \times \text{phasor OP}_1 = j^2 \times \text{phasor OP}$$
$$\text{phasor OP}_3 = j \times \text{phasor OP}_2 = j^2 \times \text{phasor OP}_1$$
$$= j^3 \times \text{phasor OP}$$

Because phasors OP and OP_2 are 180° apart,

$$\text{phasor OP}_2 = -\text{phasor OP} = -1 \times \text{phasor OP}$$
$$\text{but phasor OP}_2 = j^2 \times \text{phasor OP}$$

Therefore $j^2 = -1$.

Because the square of a real positive or negative number is always positive, j cannot be evaluated in terms of normal numbers and is therefore known as an *imaginary* quantity. As shown in Fig. 13-1, a phasor, when multiplied by j, lies along the vertical direction, which therefore can be referred to as the j or *imaginary axis*. By contrast, the horizontal direction is referred to as the *real axis*.

Because $j^3 = j^2 \times j = -1 \times j = -j$, the multiplication of a phasor by $-j$ causes the phasor to be rotated through 90° or $\pi/2$ radians in the negative or clockwise direction, leaving the magnitude of the phasor unchanged. Note that because $j^2 = -1$, $-j$ and $+j$ are reciprocals.

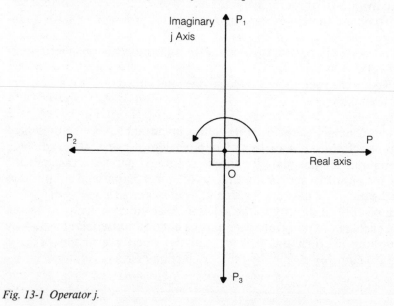

Fig. 13-1 Operator j.

RECTANGULAR AND POLAR NOTATION

Figure 13-2 represents a phasor diagram in which phasor OP = phasor r, phasor OM = phasor x and phasor OQ = phasor z.

Then phasor ON = phasor jx and because phasor OQ = phasor ON + phasor OP, then z = r + jx.

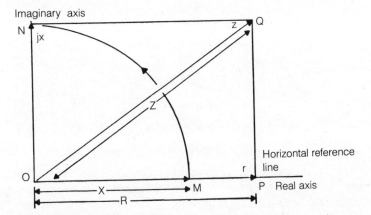

Fig. 13-2 Rectangular and polar notations.

This is known as *rectangular notation* because the phasors r and jx are 90° apart. Because phasor z is specified in terms of two phasors whose order is important (z = 2 + j3 is not the same as z = 3 + j2), this representation of a phasor is referred to as a *complex quantity.*

The polar method of denoting a phasor is in terms of its magnitude and its direction. The magnitude of z is represented by the length of the line OQ, and the direction is measured by the angle, ϕ, between OQ and the horizontal reference line. In this polar notation,

$$z = Z\underline{/\phi}$$

where Z is the magnitude of the phasor z; the value of ϕ will lie between 0° and 180°.

If the phasor lies in the lower two quadrants, the angle is negative, and then $z = Z\underline{/-\phi}$, or $Z\overline{\backslash\phi}$ where $Z\underline{/-\phi} = Z\overline{\backslash\phi}$. $Z\overline{\backslash\phi}$ should not be confused with $-Z\underline{/\phi}$ which equals $Z\underline{/\phi} + \pi$.

Conversions between rectangular and polar notations are:

If R and X are the magnitudes of the phasors r and x,

$$Z = +\sqrt{R^2 + X^2} \text{ and } \phi = \text{inv tan } \frac{X}{R}$$

Note that Z is always considered to be positive. Also

$$R = Z \cos \phi, \ X = Z \sin \phi$$

As an example, consider a resistance of 3 Ω in series with an inductive reactance of 4 Ω. Referring to Chapter 10, the phasor diagram shows the resistance phasor to be along the horizontal reference line while the reactance phasor is vertical in the "upward" direction of the j axis. The resultant of the resistance and the reactance phasors is the impedance phasor, z. In rectangular notation.

$$\text{impedance phasor, } z = 3 + j4$$
$$\text{impedance magnitude, } Z = + \sqrt{3^2 + 4^2} = 5 \ \Omega$$
$$\text{phase angle, } \phi = \text{inv tan } \frac{4}{3} = +53°8'$$

Then

$$z = 3 + j4 = 5\underline{/53°8'}$$

and

$$z = 5\underline{/53°8'} = 5 \cos 53°8' + j5 \sin 53°8'$$
$$= 3 + j4$$

Many electronic calculators have the ability to convert rapidly between rectangular and polar notation. You should familiarize yourself with the methods used by your own calculator.

Example 13-1

Convert (a) $z = 2 - j3$, $z = -5 + j2$, and $z = -4 - j7$ into polar notation, and (b) $z = 6\underline{/132°}$, $z = 8\underline{/-72°}$, $z = 12\underline{/-117°}$ into rectangular notation.

Solution

(a)
$$z = 2 - j3 = + \sqrt{2^2 + 3^2}\underline{/\text{inv tan } (-3/2)}$$
$$= 3.61\underline{/-56.3°}$$

Caution should be exercised in the determination of the angle. It is advisable to use a rough sketch to establish the quadrant in which the resultant phasor lies.

You should also remember that the acute angle obtained from the calculator always is measured with respect to the horizontal line.

$$z = -5 + j2 = + \sqrt{5^2 + 2^2} \ \underline{/\text{inv tan } (-5/2)}$$
$$= 5.39 \ \underline{/111.2°}$$
$$z = -4 - j7 = + \sqrt{4^2 + 7^2} \ \underline{/\text{inv tan } (-7/-4)}$$
$$= 8.06\underline{/-119.7°}$$

(b)
$$z = 6\underline{/132°} = 6 \cos 132° + j6 \sin 132°$$
$$= -4.01 + j4.46$$
$$z = 8\underline{/-72°} = 8 \cos (-72°) + j8 \sin (-72°)$$
$$= 2.47 - j7.61$$

$$z = 12\underline{/-117°} = 12 \cos(-117°) + j12 \sin(-117°)$$
$$= -5.45 - j10.7$$

RULES OF COMPLEX ALGEBRA

1. Equating real and imaginary parts

Let $z_1 = R_1 + jX$ and $z_2 = R_2 + jX_2$.
If $z_1 = z_2$, $R_1 + jX_1 = R_2 + jX_2$.
Then $R_1 = R_2$ and $X_1 = X_2$.

2. Addition and subtraction of phasors

If $z_1 = R_1 + jX_1$ and $z_2 = R_2 + jX_2$, then
$$z_1 + z_2 = (R_1 + jX_1) + (R_2 + jX_2) = (R_1 + R_2) + j(X_1 + X_2)$$
$$\text{and } z_1 - z_2 = (R_1 - R_2) + j(X_1 - X_2).$$

Rectangular rather than polar notation is used when adding or subtracting phasors.

3. Multiplication of phasors

If $z_1 = R_1 + jX_1$ and $z_2 = R_2 + jX_2$, then
$$z_1 z_2 = (R_1 + jX_1)(R_2 + jX_2)$$
$$= R_1 R_2 + jR_1 X_2 + jR_2 X_1 + j^2 X_1 X_2$$
$$= (R_1 R_2 - X_1 X_2) + j(R_1 X_2 + R_2 X_1)$$

because $j^2 = -1$.

Using polar notation, $z_1 = Z_1\underline{/\phi_1}$
$$\text{and } z_2 = Z_2\underline{/\phi_2}$$

Then

$$z_1 z_2 = Z_1(\cos \phi_1 + j \sin \phi_1) \times Z_2(\cos \phi_2 + j \sin \phi_2)$$
$$= Z_1 Z_2[(\cos \phi_1 \cos \phi_2 - \sin \phi_1 \sin \phi_2)$$
$$+ j(\sin \phi_1 \cos \phi_2 + \cos \phi_1 \sin \phi_2)]$$
$$= Z_1 Z_2[\cos(\phi_1 + \phi_2) + j \sin(\phi_1 + \phi_2)]$$
$$= Z_1 Z_2\underline{/\phi_1 + \phi_2}$$

When multiplying phasors, the magnitudes are multiplied, but the angles are added. It is preferable to use polar notation when multiplying or dividing phasors.

If $z = Z\underline{/\phi}$, $z^2 = Z\underline{/\phi} \times Z\underline{/\phi} = Z^2\underline{/2\phi}$.

Therefore in squaring a phasor, the magnitude must be squared, but the angle is doubled. It follows that in taking the square root of a phasor, you must obtain the square root of the magnitude, but the angle is halved.

$$\sqrt{z} = \sqrt{Z\underline{/\phi}} = \sqrt{Z}\underline{/\phi/2} = Z^{1/2}\underline{/\phi/2}.$$

4. Division of Phasors

If $z_1 = R_1 + jX_1$, $z_2 = R_2 + jX_2$.

$$\frac{z_1}{z_2} = \frac{R_1 + jX_1}{R_2 + jX_2}$$

In order to separate out the real and imaginary parts of z_1/z_2, it is necessary to eliminate j from the denominator. This is done by *rationalization* which means multiplying both numerator and denominator by the conjugate of the denominator. The *conjugate* of a phasor is that phasor which has the same magnitude but whose angle is opposite in sign, although equal in magnitude. Therefore the conjugate of $Z\underline{/\phi}$ is $Z\underline{/-\phi}$, and the conjugate of $R + jX$ is $R - jX$. Then:

$$\frac{z_1}{z_2} = \frac{(R_1 + jX_1)(R_2 - jX_2)}{(R_2 + jX_2)(R_2 - jX_2)}$$

$$= \frac{(R_1 R_2 + X_1 X_2)}{R_2{}^2 + X_2{}^2} + \frac{j(X_1 R_2 - X_2 R_1)}{R_2{}^2 + X_2{}^2}$$

In polar form, $z_1 = Z_1\underline{/\phi_1}, = Z_2\underline{/\phi_2}$ so that

$$\frac{z_1}{z_2} = \frac{Z_1(\cos \phi_1 + j \sin \phi_1)}{Z_2(\cos \phi_2 + j \sin \phi_2)}$$

$$= \frac{Z_1}{Z_2} \times \left[\frac{(\cos \phi_1 + j \sin \phi_1)(\cos \phi_2 - j \sin \phi_2)}{(\cos \phi_2 + j \sin \phi_2)(\cos \phi_2 - j \sin \phi_2)} \right]$$

$$= \frac{Z_1}{Z_2} \times \left[\frac{(\cos \phi_1 \cos \phi_2 + \sin \phi_1 \sin \phi_2) + j(\sin \phi_1 \cos \phi_2 - \cos \phi_1 \sin}{\cos^2 \phi_2 + \sin^2 \phi_2} \right.$$

$$= \frac{Z_1}{Z_2} \times \left[\frac{\cos (\phi_1 - \phi_2) + j \sin (\phi_1 - \phi_2)}{1} \right]$$

$$= \frac{Z_1}{Z_2}\underline{/\phi_1 - \phi_2}$$

When dividing phasors, the magnitudes are divided, but the angles are subtracted. Note the special case of the reciprocal, $z = Z\underline{/\phi^\circ}$.

$$\frac{1}{z} = \frac{1\underline{/0^\circ}}{Z\underline{/\phi^\circ}} = \frac{1}{Z}\underline{/-\phi^\circ}$$

See Figs. 13-3A, 13-3B and 13-3C.

Example 13.2

If $z_1 = 3 + j5$ and $z_2 = 4 - j7$, find the values of $z_1 + z_2$, $z_1 - z_2$, $z_1 \times z_2$, z_1/z_2, $z_1{}^2$, $\sqrt{z_2}$, $1/z_1$.

Solution

$$z_1 = 3 + j5 = \sqrt{3^2 + 5^2}\underline{/\text{inv tan } 5/3} = 5.83\underline{/59.0^\circ}.$$

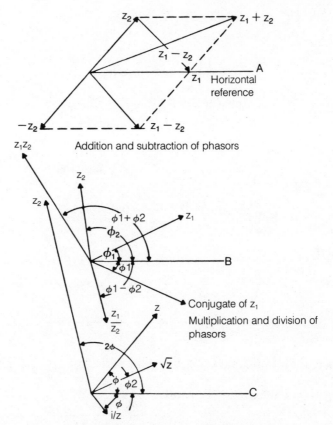

Addition and subtraction of phasors

Multiplication and division of phasors

Conjugate of z_1

Square. Square root and reciprocal of a phasor

Fig. 13-3 Phasor illustrations of the rules of complex algebra.

$$z_2 = 4 - j7 = \sqrt{4^2 + 7^2}\underline{/\text{inv tan}\ (-7/4)} = 8.06\underline{/-60.3°}.$$

$$z_1 + z_2 = (3 + j5) + (4 - j7) = 7 - j2$$
$$= \sqrt{7^2 + 2^2}\underline{/\text{inv tan}\ (-2/7)}$$
$$= 7.28 - \underline{/16.0°}$$

$$z_1 - z_2 = (3 + j5) - (4 - j7) = -1 + j12$$
$$= \sqrt{1^2 + 12^2}\underline{/\text{inv tan}\ 12/1}$$
$$= 12.04\underline{/94.8°}$$

$$z_1 z_2 = 5.83\underline{/59.0°} \times 8.06\underline{/-60.3°} = 47.0\ \underline{/-1.3°}$$

$$z_1/z_2 = 5.83\underline{/59.°}\ /8.06\underline{/-60.3°} = 0.723\underline{/119.3°}$$

$$z_1{}^2 = 5.8^2\underline{/2 \times 59.0°} = 33.64\underline{/118°}$$

$$\sqrt{z_2} = \sqrt{8.06}\underline{/-60.3°/2} = 2.84\underline{/-30.15°}$$

$$\frac{1}{z_1} = z_1{}^{-1} = \frac{1\underline{/0°}}{5.83\underline{/59.0°}} = 0.172\underline{/-59.0°}$$

ANALYSIS OF AC CIRCUITS

Resistance

$$Z = \frac{E_{rms}}{I_{rms}} = R, \phi = 0°$$

$$\text{phasor } z = R\underline{/0°} = R + j0$$

Inductance

$$Z = \frac{E_{rms}}{I_{rms}} = \omega L = 2\pi fL = X_L, \phi = +90°$$

$$\text{phasor } z = \omega L\underline{\left/\frac{\pi}{2}\right.} = 0 + j\omega L$$

Capacitance

$$Z = \frac{E_{rms}}{I_{rms}} = \frac{1}{\omega C} = \frac{1}{2\pi fC} = X_C, \phi = -90°$$

$$\text{phasor } z = \frac{1}{\omega C}\underline{\left/-\frac{\pi}{2}\right.} = 0 - \frac{j}{\omega C} = 0 + \frac{1}{j\omega C}$$

Resistance, Inductance and Capacitance in Series

$$\text{phasor } z = R + j\omega L - \frac{j}{\omega C}$$

$$\frac{E_{rms}}{I_{rms}} = Z = \sqrt{R^2 + \left(\omega L - \frac{1}{\omega C}\right)^2}$$

$$\text{phase angle, } \phi = \text{inv tan} \left(\frac{\omega L - \dfrac{1}{\omega C}}{R}\right)$$

Resistance, Inductance and Capacitance in Parallel

$$\frac{1}{z} = \frac{1}{R} + \frac{1}{j\omega L} + \frac{1}{-j/\omega C} = \frac{1}{R} + \frac{1}{j\omega L} + j\omega C$$

or

$$y = G - jB_L + jB_C$$

Then

$$Y = \frac{1}{Z} = \sqrt{G^2 + (B_C - B_L)^2}$$

and

$$\phi = \text{inv tan} \left(\frac{B_C - B_L}{G} \right)$$

The use of operator j in the analysis of more complex ac circuits is illustrated in the following examples.

Example 13-3

In the series-parallel circuit of Fig. 13-4, express the total impedance phasor and the supply current in their polar forms.

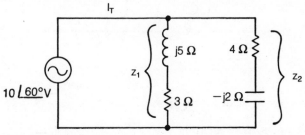

Fig. 13-4 Circuit for example 13-3.

Solution

impedances $z_1 = 3 + j5$ and $z_2 = 4 - j2$.

Using the product-over-sum formula for the total impedance phasor,

$$z_T = \frac{z_1 z_2}{z_1 + z_2} = \frac{(3 + j5)(4 - j2)}{(3 + j5) + (4 - j2)} = \frac{(3 + j5)(4 - j2)}{(7 + j3)}$$

Because we are now faced with multiplication and division of phasors, it is more convenient to convert to the polar form. Then

$$z_T = \frac{\sqrt{3^2 + 5^2}\underline{/\text{inv tan } 5/3} \times \sqrt{4^2 + 2^2}\underline{/\text{inv } (-2/4)}}{\sqrt{7^2 + 3^2}\underline{/\text{inv tan } 3/7}}$$

$$= \frac{5.83\underline{/59.04°} \times 4.47\underline{/-26.57°}}{7.62\underline{/23.20°}}$$

$$= \frac{5.83 \times 4.47}{7.62} \underline{/59.04° - 26.57° - 23.20°}$$

$$= 3.42\underline{/9.27°} \ \Omega$$

Supply current

$$I_T = \frac{10\underline{/60°}}{3.42\underline{/9.27°}} = 2.92\underline{/50.73°} \text{ A.}$$

Example 13-4

In Fig. 13-5, calculate the value of the total impedance phasor, z_t

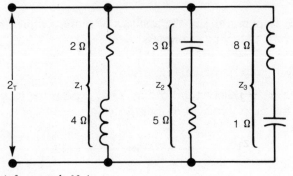

Fig. 13-5 Circuit for example 13-4.

Solution

impedances $z_1 = 2 + j4 \, \Omega$, $z_2 = 5 - j3 \, \Omega$, $z_3 = j8 - j7 \, \Omega$

From the reciprocal formula,

$$\frac{1}{z_t} = \frac{1}{z_1} + \frac{1}{z_2} + \frac{1}{z_3}$$

$$= \frac{1}{2 + j4} + \frac{1}{5 - j3} + \frac{1}{j8 - j1}$$

Using the rationalization process,

$$\frac{1}{z_t} = \frac{2 - j4}{2^2 + 4^2} + \frac{5 + j3}{5^2 + 3^2} + \frac{1}{j7}$$

$$= \frac{2 - j4}{20} + \frac{5 + j3}{34} - j0.143$$

$$\frac{1}{z_t} = 0.1 - j0.2 + 0.147 + j0.088 - j0.143$$

$$= 0.188 - j0.255 = 0.317\underline{/-53.6°} \text{ S}$$

Then,

$$\text{total impedance, } z_t = \frac{1}{0.317\underline{/-53.6°}} = 3.15\underline{/53.6°} \, \Omega$$

CHAPTER SUMMARY

☐ Rectangular to Polar Notation

$$z = R + jX = Z\underline{/\phi}, \text{ where } Z = \sqrt{R^2 + X^2} \text{ and } \phi = \text{inv tan } \frac{X}{R}$$

☐ Polar to Rectangular Notation

$z = Z\underline{/\phi} = Z \cos \phi + jZ \sin \phi = R + jX$ where $R = Z \cos \phi$ and $X = Z \sin \phi$.

☐ Rules of Complex Algebra

$$z_1 = Z_1\underline{/\phi_1} = R_1 + jX_1, z_2 = Z_2\underline{/\phi_2} = R_2 + jX_2$$

Equating "real" and "imaginary" parts:

$$\text{If } R_1 + jX_1 = R_2 + jX_2, R_1 = R_2, \text{ and } X_1 = X_2.$$

Addition:

$$z_1 + z_2 = (R_1 + R_2) + j(X_1 + X_2).$$

Subtraction:

$$z_1 - z_2 = (R_1 - R_2) + j(X_1 - X_2).$$

Multiplication:

$$z_1 z_2 = Z_1 Z_2 \underline{/\phi_1 + \phi_2}$$

Square:

$$z_1^2 = Z_1^2 \underline{/2\phi_1}$$

Square root:

$$\sqrt{z_1} = \sqrt{Z_1} \underline{/\phi_1/2}$$

Division:

$$\frac{z_1}{z_2} = \frac{Z_1}{Z_2} \underline{/\phi_1 - \phi_2}$$

Reciprocal:

$$\frac{1}{z_1 \underline{/\phi_1}} = \frac{1}{Z_1} \underline{/-\phi_1}$$

Conjugates:

the conjugate of $z = Z\underline{/\phi} = R + jX$ is $z' = Z\underline{/-\phi} = R - jX$.

Rationalization:

$$\text{to rationalize } \quad \frac{A + jB}{C + jD}$$

multiply the numerator and the denominator by the denominator's conjugate, $C - jD$.

14
Filters

As we cover the subject of filters, you will understand:

☐ The purpose of filters and their use of reactive elements to obtain the required attenuation-versus-frequency response curves.
☐ How to compare the T and π filter sections.
☐ The operation of the prototype or constant-k low-pass filter.
☐ The operation of constant-k high-pass, band-pass, and band-stop filters.
☐ Why the m-derived filter is superior to the constant-k variety.
☐ The operation of m-derived low-pass, high-pass, band-pass, and band-stop filters.

PRINCIPLES OF FILTERS

A network that is designed to attenuate certain frequencies but pass others is called a *filter*. A filter therefore possesses at least one *pass band,* a band of frequencies for which the attenuation theoretically is zero, and at least one stop band, a band of frequencies for which the attenuation in decibels is finite. The frequencies that separate the various pass and stop bands are called the *cutoff values.*

An important characteristic of all filters is that they are constructed from purely reactive elements because the attenuation otherwise could never become zero. Certain networks containing resistances and reactances have a filtering action, but they are not filters in the true sense. An *ideal filter* presents zero attenuation in the pass band and infinite attenuation in the stop band. For example, an ideal low-pass filter has an attenuation-versus-frequency characteristic as shown in Fig. 14-1.

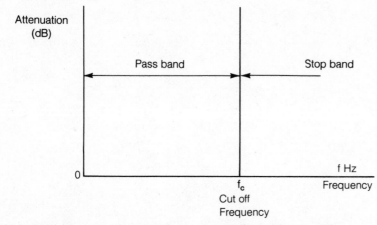

Fig. 14-1 Attenuation versus frequency characteristic of an ideal low-pass filter.

The ideal characteristic cannot be achieved in practice because of filter losses, such as are caused by the inductor's effective resistance. Furthermore, there can be a mismatch at certain frequencies between the filter and its terminating load. This also affects the shape of the attenuation characteristic.

T AND π FILTER SECTIONS

As shown in Fig. 14-2, a filter can be regarded as being composed of a number of repeating T and π sections. Z_1 is referred to as the filter's series impedance while Z_2 is the shunt impedance.

Consider a T section that has to be matched to a load of value Z_o (Fig. 14-3).
Then

$$Z_o = \frac{Z_1}{2} + \frac{\left(Z_o + \frac{Z_1}{2}\right) \times Z_2}{Z_o + \frac{Z_1}{2} + Z_2}$$

which leads to

$$Z_o = \sqrt{\frac{Z_1^2}{4} + Z_1 Z_2} \ \Omega$$

Provided this condition is fulfilled, any number of T sections can be added without disturbing the matching condition between the filter and its terminating load.

Constant-k Low-Pass Filter

The constant-k variety is a simple or prototype filter for which the k factor ($= \sqrt{z_1 z_2}$) is a fixed value that is independent of the frequency.

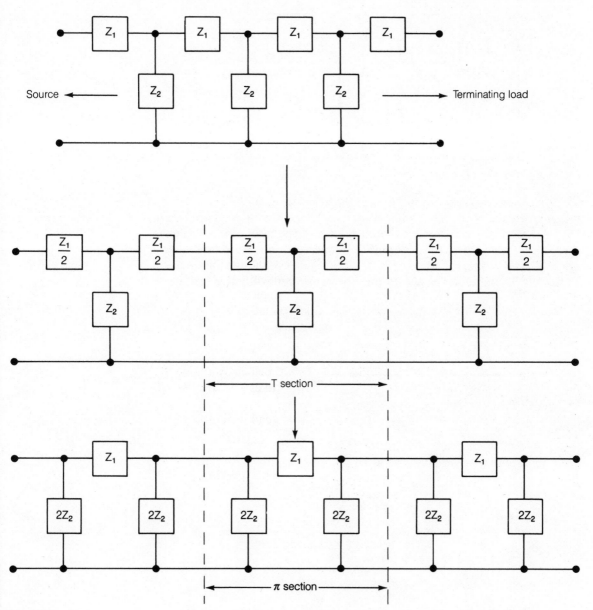

Fig. 14-2

For example, with either π or T sections, if Z_1 is an inductor and Z_2 is a capacitor (or vice-versa), the k factor is given by

$$k = \sqrt{2\pi fL \times \frac{1}{2\pi fC}} = \sqrt{\frac{L}{C}}$$

which is a constant and is independent of the frequency.

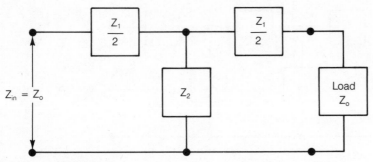

Fig. 14-3 The matched T section.

An example of a constant-k filter is the low-pass circuit of Fig. 14-4. At low frequencies, V_L is small while V_C exceeds V_i (because V_L and V_C are 180° out of phase). It is possible therefore for V_o to equal V_1 and this means theoretically zero attenuation. At high frequencies, V_L is large so that V_C and hence V_o are low and the attenuation is high. Figure 14-5 shows the resulting characteristic and it can be shown that the cutoff frequency is given by

$$\text{cutoff frequency, } f_C = \frac{1}{\pi \sqrt{LC}} \text{ Hz}$$

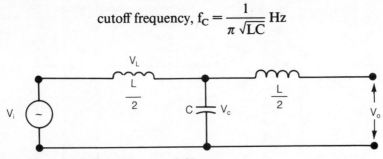

Fig. 14-4 T-section of a low-pass constant-k filter.

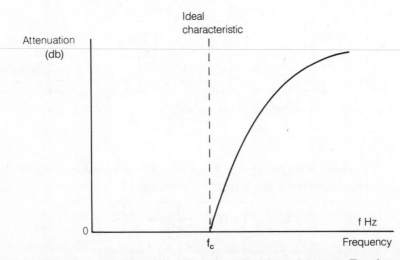

Fig. 14-5 Attenuation versus frequency characteristic of a constant-k low-pass T section.

High-Pass Filter

With this type of filter L and C are interchanged and the basic T section appears in Fig. 14-6.

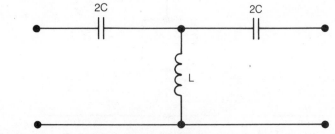

Fig. 14-6 T section of a high-pass constant-k filter.

The cutoff frequency is given by

$$f_C = \frac{1}{4 \pi \sqrt{LC}} \text{ Hz}$$

and the associated attenuation versus frequency curve appears in Fig. 14-7.

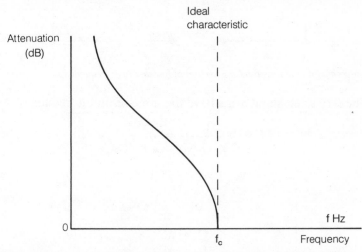

Fig. 14-7 Attenuation versus frequency characteristic of a constant-k high pass filter.

Band-Pass Filter

This type of filter must have two cutoff values because it is designed to pass a band of frequencies with zero attenuation. Its basic band-pass T section appears in Fig. 14-8.

At low frequencies, the section behaves as a high-pass filter (Fig. 14-9A) while at the high frequencies the section appears to be a low-pass filter (Fig. 14-9B). Consequently there is a high degree of attenuation for both low and high frequencies.

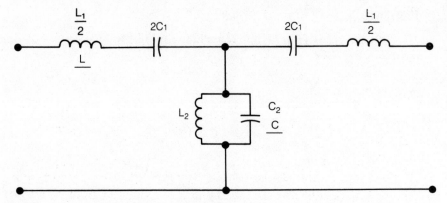

Fig. 14-8 *T section of a band-pass constant-k filter.*

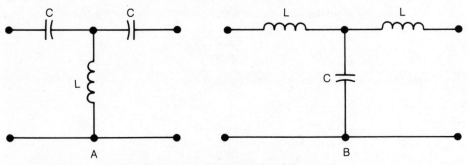

Fig. 14-9 *Behavior of the band-pass T section at low and high frequencies.*

The series and shunt arms have the same resonant frequency

$$f_o = \frac{1}{2\pi\sqrt{L_1 C_1}} = \frac{1}{2\pi\sqrt{L_2 C_2}}$$

Let

$$L_1 = \frac{L}{n}$$
$$C_1 = nC$$
$$L_2 = nL$$
$$C_2 = \frac{C}{n}$$

Then

$$z_1 = j\left(\frac{\omega L}{2n} - \frac{1}{n\omega C}\right) = \frac{j}{n\omega C}(\omega^2 LC - 1)$$

and

$$z_2 = \frac{j\omega nL}{1 - \omega nL \times \dfrac{\omega C}{n}} = \frac{j\omega nL}{1 - \omega^2 LC}$$

then

$$k = \sqrt{z_1 z_2} = \sqrt{\frac{j}{\omega nC}(\omega^2 LC - 1) \times \frac{j\omega nL}{1 - \omega^2 LC}} = \sqrt{\frac{L}{C}}$$

Consequently k is again a constant that is independent of frequency.

For both high and low frequencies, the reactance of the series arm is high and that of the shunt arm is low so that there is a degree of finite attenuation. However, at the frequency f_o, the impedance of the series arm is theoretically zero while that of the shunt arm is infinite; consequently there is zero attenuation. The resulting attenuation-versus-frequency curve appears in Fig. 14-10.

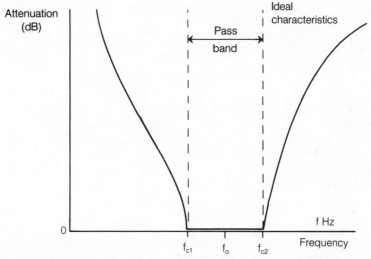

Fig. 14-10 Attenuation versus frequency characteristic of a constant-k band-pass filter.

Band-Stop Filter

This type of filter is designed to "stop" a band of frequencies and therefore must also possess two cutoff values. Compared with the band-pass type, the series and shunt arms are interchanged, and therefore the basic T section is as shown in Fig. 14-11. Its attenuation-versus-frequency characteristic appears in Fig. 14-12.

INTRODUCTION TO M-DERIVED FILTERS

The main problem with constant-k filters is that the attenuation does not rise or fall very rapidly near the cutoff frequencies. It would appear at first

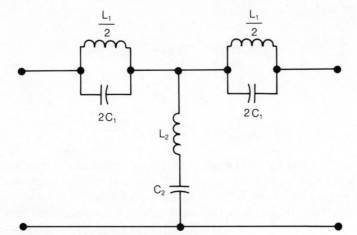

Fig. 14-11 T section of a band-stop constant-k filter.

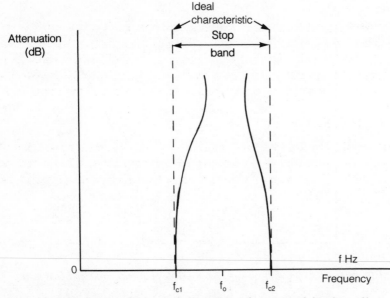

Fig. 14-12 Attenuation versus frequency characteristics of a constant-k band stop filter.

glance that this could be solved by connecting a number of identical T sections together. This would not disturb the matching of the filter and while the attenuation over the pass band theoretically would still be zero, the attenuation at other frequencies would be greatly increased. Unfortunately, because of losses in the components, the attenuation in the pass band of a practical filter is not zero, and the attenuation-versus-frequency curve becomes badly "rounded" near the cutoff value (Fig. 14-13).

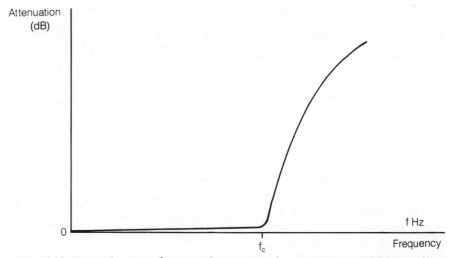

Fig. 14-13 Attenuation versus frequency characteristic of a practical constant-k low-pass filter.

It is therefore required to design a new filter section that has the same cutoff frequency as that of the constant-k type and is matched to its terminating load, but it will have an improved (i.e., sharper) attenuation-versus-frequency characteristic. This is the purpose of the m-derived filter in which m is a mathematical parameter whose value lies between zero and unity.

M-Derived T Section

Consider any T section in which the Z_1 series impedance of the prototype filter (Fig. 14-14A) is replaced by mZ_1 (in which m is some constant) and the shunt impedance, Z_2, by $Z_2{}^1$. The result is shown in Fig. 14-14B. Because the matched load Z_o for the prototype filter is

$$\sqrt{\frac{Z_1{}^2}{4} + Z_1 Z_2}\ \Omega,$$

the corresponding Z_o for the m-derived section is

$$\sqrt{\frac{m^2 Z_1{}^2}{4} + m Z_1 Z_2{}^1}\ \Omega$$

If the two sections are equivalent for matching purposes,

$$\frac{Z_1{}^2}{4} + Z_1 Z_2 = \frac{m^2 Z_1{}^2}{4} + m Z_1 Z_2{}'$$

This yields

$$Z_2{}' = \frac{Z_2}{m} + \frac{(1-m^2)Z_1}{4m}$$

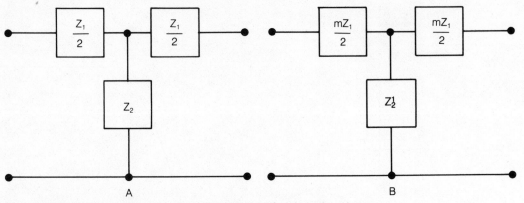

Fig. 14-14 Comparison between constant-k (prototype) and m-derived T sections.

Therefore the shunt impedance, Z_2', consists of an impedance Z_2/m in series with another impedance

$$\frac{Z_1 (1 - m^2)}{4m}$$

that can be constructed provided $0 < m < 1$. The complete m-derived T section appears in Fig. 14-15.

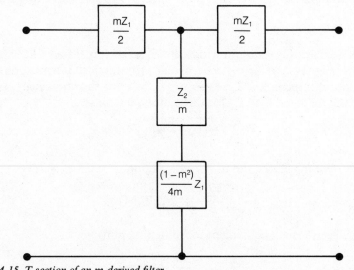

Fig. 14-15 T section of an m-derived filter.

Low-Pass M-Derived Filter

The related T section for a low-pass m-derived filter appears in Fig. 14-16. At low frequencies, the combination of the reactance mC in series with the reactance $(1 - m^2)$ L/4m behaves capacitively and therefore provides a pass band. The attenuation band begins at the cutoff frequency, f_c,

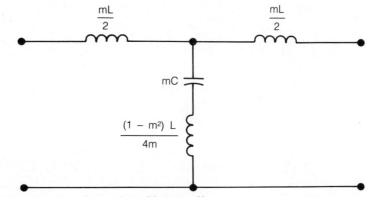

Fig. 14-16 T section of an m-derived low-pass filter.

but above this value, the reactance mC resonates with the reactance $(1 - m^2)$ L/4m at another frequency, f_∞, and therefore a short circuit theoretically exists in the shunt arm, and infinite attenuation is the result. For still higher frequencies, the shunt arm behaves inductively, and there is a high degree of attenuation (Fig. 14-17).

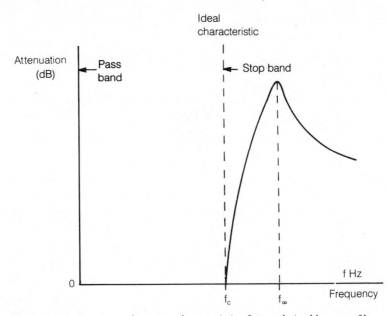

Fig. 14-17 Attenuation versus frequency characteristic of an m-derived low-pass filter.

In Figure 14-17, the frequency, f_∞, is given by

$$\frac{f_\infty{}^2}{4 \pi^2} = \omega_\infty{}^2$$

$$= \frac{1}{\dfrac{(1-m^2)L}{4m} \times mC}$$

$$= \frac{4}{LC\,(1-m^2)}.$$

$$= \frac{\omega_c^2}{1-m^2}$$

This means that f_∞ is greater than f_c.

The shape of the attenuation-versus-frequency characteristic is much "sharper" near the cutoff value, and therefore the m-derived filter is a considerable improvement over the LC prototype variety.

HIGH PASS, BAND BASS, AND BAND STOP M-DERIVED FILTERS

The T sections and the corresponding attenuation-versus-frequency curves for the m-derived high-pass, band-pass, and band-stop filters appear in Fig. 14-18.

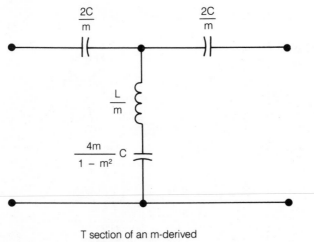

T section of an m-derived
high-pass filter

Fig. 14-18 High-pass, band-pass, and band-stop m-derived filters.

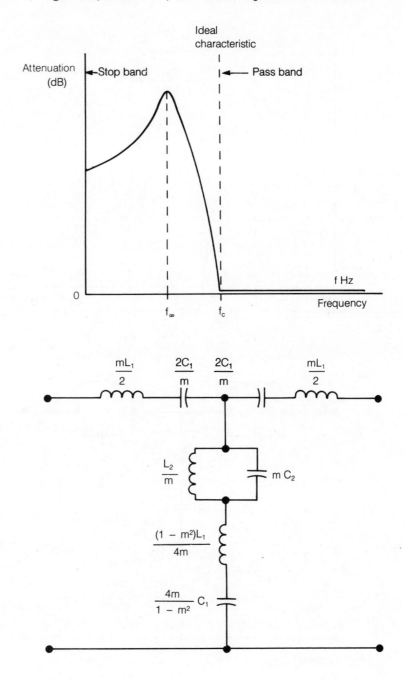

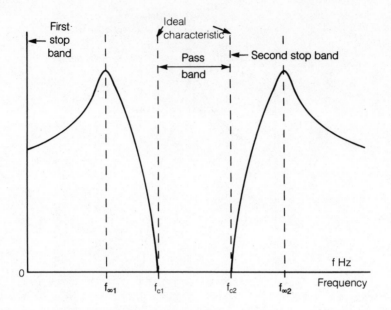

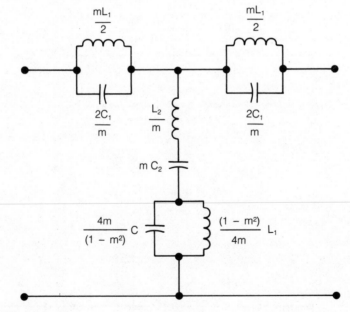

Fig. 14-18 Continued.

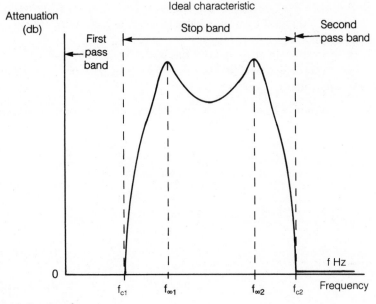

Fig. 14-18 Continued.

CHAPTER SUMMARY

☐ Filters in General

1. A filter is a reactive network that is designed to stop certain frequencies but to pass other frequencies.
2. A filter's k factor is the value of $\sqrt{z_1 z_2}$,, where z_1 is the series impedance and z_2 is the shunt impedance.
3. For a T section to be matched to a load of value, z_o, it is necessary that

$$z_o = \sqrt{z_1 z_2 + \frac{z_1^2}{4}} \; \Omega$$

☐ Constant-k Filters

4. A constant-k (prototype) low-pass filter has one low frequency pass band and one high-frequency stop band. Cutoff frequency, $f_c = 1/\pi \sqrt{LC}$ Hz.
5. A constant-k high-pass filter has one low frequency stop band and one high-frequency pass band. Cutoff frequency is

$$f_c = \frac{1}{4 \pi \sqrt{LC}} \; Hz$$

6. A constant-k band-pass filter has both low- and high-frequency stop bands separated by a pass band. Consequently there are two cutoff frequencies.

7. A constant-k band-stop filter has both low- and high-frequency pass bands separated by a stop band with a cutoff frequency on either side.

□ M-Derived Filters

8. An m-derived filter has sharper characteristics than its constant-k counterpart. The value of the mathematical parameter, m, lies between zero and unity.

15
Mutually
Coupled Circuits

As WE INVESTIGATE the subject of mutually coupled circuits, you will learn:

☐ That the property of mutual inductance is associated with the flux of one coil linking with the turns of a second coil.

☐ About the coupling factor relationship between the mutual inductance and the self-inductances of the two coils.

☐ How to calculate the value of the total effective inductance for two mutually coupled coils in series aiding or series opposing.

☐ How to calculate the value of the total effective inductance for two mutually coupled coils in parallel opposing.

☐ The principles of the power transformer, which provides the ability to step-up or step-down an alternating voltage.

☐ About the losses that occur in a practical power transformer and its efficiency.

☐ How the reflected resistance represents the effect of a secondary load on the primary source.

Previous chapters discussed inductors having a magnetic flux linked with their own turns and therefore possessing the property of self-inductance. We are now going to consider the case in which the magnetic flux associated with one coil links with the turns of another coil. Such inductors are said to be *magnetically, inductively, mutually,* or *transformer coupled* and possess the property of mutual inductance, M.

MUTUAL INDUCTANCE

The alternating current i_1 in the coil 1 of Fig. 15-1 creates an alternating magnetic flux, part of which only links coil 1 (this is referred to as the leakage

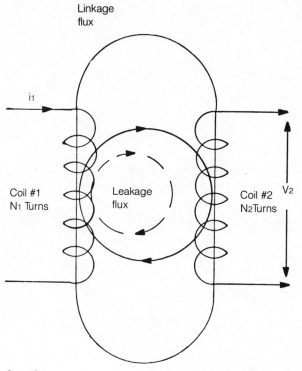

Fig. 15-1 Mutual coupling.

flux), and the remainder links coil 2. This causes an induced voltage, v_2, whose magnitude, in part, depends on i_1 and the value of the mutual inductance between the coils. Mutual inductance, M, like self-inductance, is measured in henrys. The mutual inductance is 1 henry if, when the current i_1 is instantaneously changing at the rate of 1 ampere per second, the induced voltage, v_2 is 1 volt. The factors that determine the value of the mutual inductance include the number of turns for N_1 and N_2, the cross-sectional area of the coils, their separation, the orientation of their axes, and the nature of their cores.

Induced voltage, $v_2 = M \times$ rate of change in i_1. In terms of ac sinewave values, $V_{2rms} = 2\pi fMI_{1rms}$, where f is the frequency of the current i_1 (compare $V_{Lrms} = 2\pi fLI_{rms}$ for the property of self-inductance). Provided the two coils are wound in the same sense, v_2 lags i_1 by 90°, but if the coils are in the opposite sense, v_2 leads i_i by 90°. Note that the property of mutual inductance is reversible so that if the same rate of change of the current, i_1, is flowing in coil 2, then the voltage induced in coil 1 is $v_1 = M \times$ rate of change in i_1.

In the extreme case in which the two coils are tightly wound, one on top of the other with a common soft-iron core, the leakage flux is extremely small and can be neglected. Assuming perfect flux linkage between the coils (corresponding to zero leakage flux), the mutual inductance is

$$M = \frac{\mu_0 \mu_r N_1 N_2 A}{\ell} \text{ henrys}$$

where

A is the cross-sectional area of each of the coils in square meters
μ_r is the relative permeability of the soft iron
ℓ is the length of the coils in meters
$\mu_0 = 4\,\pi \times 10^{-7}$ SI units.

Because the self-inductance of the coil 1 is

$$L_1 = \frac{\mu_0 \mu_r N_1{}^2 A}{\ell} \text{ henrys}$$

and, in a similar way,

$$L_2 = \frac{\mu_0 \mu_r N_2{}^2 A}{\ell} \text{ henrys}$$

Then

$$M^2 = L_1 L_2 \quad \text{and} \quad M = \sqrt{L_1 L_2} \text{ henrys}$$

THE COUPLING FACTOR

If the leakage flux is not negligible, then only a fraction, k, of the total flux links the two coils. This fraction k, which cannot exceed 1, is called the *coefficient of coupling,* or *coupling factor;* its value can be close to 1 if a common soft-iron core is used for the two coils, but can be very small (less than 0.01) with an air core and the coils widely separated.

It can be shown that

$$k = \frac{M}{\sqrt{L_1 L_2}} \quad \text{or} \quad M = k \times \sqrt{L_1 L_2}$$

Then

$$L_1 = \frac{M^2}{k^2 L_2}$$

and

$$L_2 = \frac{M^2}{k^2 L_1}$$

Also

$$M = kL_1 \times \frac{N_2}{N_1} = kL_2 \times \frac{N_1}{N_2}$$

If L_1 and L_2 are equal to L,

$$k = \frac{M}{L}$$

and

$$M = kL$$

Note that if a steady direct current flows through coil 1, the linkage flux is constant in magnitude and direction so that the voltage induced in coil 2 is zero.

Mutually Coupled Coils in Series

If two mutually coupled coils having a common axis are connected in series, their individual fluxes can aid or oppose, depending on the sense in which the coils are wound.

The direction of the arrows, which indicates current flow, clearly demonstrates that in Fig. 15-2A the individual fluxes surrounding L_1 and L_2 are aiding while in Fig. 15-2B, they are opposing.

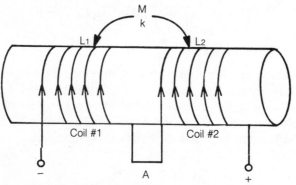

Fig. 15-2 A Coils connected in series-aiding.

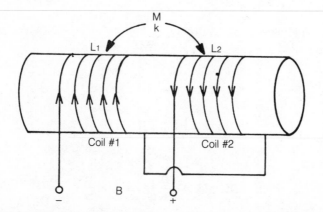

Fig. 15-2 B Coils connected in series-opposing.

First consider the case in which the fluxes are aiding, as shown in Fig. 15-3. Let the rate of change of current through the series circuit be I/t amperes per second. The voltage induced in coil 1 is $L_1 I/t$ volts because of self-inductance and MI/t volts because of the flux surrounding coil 2 and linking partially with coil 1. Likewise, the voltage induced in coil 2 is $L_2 I/t$ from self-inductance and MI/t from mutual inductance. Because the fluxes are aiding, the total voltage across the series circuit is

$$(L_1 I/t + MI/t) + (L_2 I/t + MI/t) = (L_1 + L_2 + 2M)I/t$$

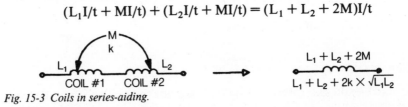

Fig. 15-3 Coils in series-aiding.

Then the total inductance, L_T is $L_1 + L_2 + 2M$, as shown in the equivalent circuit of Fig. 15-3.

If the fluxes are opposing, the sign of M reverses, and the total inductance becomes $L_T = L_1 + L_2 - 2M$.

The equations are
Series-aiding coils:

$$L_T^+ = L_1 + L_2 + 2M = L_1 + L_2 + 2k \times \sqrt{L_1 L_2}$$

Series-opposing coils:

$$L_T^- = L_1 + L_2 - 2M = L_1 + L_2 - 2k \times \sqrt{L_1 L_2}$$

Therefore,

$$L_1 + L_2 = \frac{L_T^+ + L_T^-}{2}$$

and

$$M = \frac{L_T^+ - L_T^-}{4}$$

The positive sign used with L_T means aiding, the negative sign means opposing.

If L_1 and L_2 are each equal to L, the equations become
Series-aiding coils:

$$L_T^+ = 2(L + M) = 2L(1 + k)$$

Series-opposing coils:

$$L_T^- = 2(L - M) = 2L(1 - k)$$

Then

$$L = \frac{L_T^+ + L_T^-}{4}, \ M = \frac{L_T^+ - L_T^-}{4}, \ k = \frac{L_T^+ - L_T^-}{L_T^+ + L_T^-}$$

Mutually Coupled Coils in Parallel

Fig. 15-4 represents two mutually coupled coils in parallel. If the sense of the inductors is such that the coils are in parallel-aiding, the total inductance, L_T^+, between points A and B is

$$L_T^+ = \frac{L_1 L_2 - M^2}{L_1 + L_2 - 2M}$$

However, if the coils are connected in parallel-opposing, the sign of M is reversed and the total inductance is

$$L_T^- = \frac{L_1 L_2 - M^2}{L_1 + L_2 + 2M}$$

(The above expressions for L_T^+ and L_1^- are not the same as those derived from

$$\frac{1}{L_T} = \frac{1}{L_1 \pm M} + \frac{1}{L_2 \pm M}$$

which is shown erroneously in some textbooks).

Note that if $M = 0$, both equations become

$$L_T = \frac{L_1 L_2}{L_1 + L_2}$$

which is the familiar product-over-sum formula for self-inductances.

If L_1 and L_2 are each equal to L, the equations become

$$L_T^+ = \frac{L^2 - M^2}{2(L - M)} = \frac{L + M}{2}$$

and

$$L_T^- = \frac{L - M}{2}$$

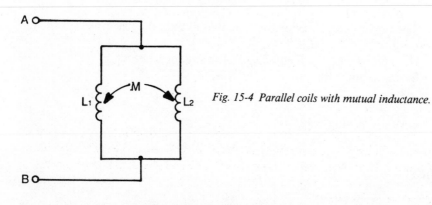

Fig. 15-4 Parallel coils with mutual inductance.

Example 15-1

Two mutually coupled coils are joined in a series-aiding arrangement. If the self-inductances are 0.75 H, 0.6 H, and the coupling factor is 0.8, find the total equivalent inductance. If one of the coils is now reversed without changing the coupling factor, what is the new value of the total equivalent inductance?

Solution

$$\text{mutual inductance, } M = k \times \sqrt{L_1 L_2} = 0.8 \times \sqrt{0.75 \times 0.6}$$
$$= 0.537 \text{ H}$$
$$\text{total equivalent inductance, } L_T{}^+ = L_1 + L_2 + 2M$$
$$= 2.42 \text{ H}$$

If one of the coils is reversed, the connection is now series-opposing.

$$\text{total equivalent inductance, } L_T{}^- = L_1 + L_2 - 2M$$
$$= 0.276 \text{ H}$$

Example 15-2

Two coils whose self-inductances are 75 μH and 125 μH have a mutual inductance of 15 μH. What is the coupling factor? Calculate the equivalent inductance if the coils are connected in (a) series-aiding and (b) series-opposing.

Solution

$$\text{coupling factor, } k = \frac{M}{\sqrt{L_1 L_2}} = \frac{15}{\sqrt{75 \times 125}}$$
$$= 0.155$$

(a) In the series-aiding connection,

$$L_T{}^+ = L_1 + L_2 + 2M = 75 + 125 + 2 \times 15.5$$
$$= 231 \ \mu H$$

(b) In the series-opposing connection,

$$L_T{}^- = L_1 + L_2 - 2M = 75 + 125 - 2 \times 15.5$$
$$= 169 \ \mu H$$

Example 15-3

Two coils whose self-inductances are 65 mH and 85 mH are connecting in parallel-aiding, with a coupling factor of 0.35. What is the total equivalent inductance of the parallel combination? If one of the coils is now reversed without changing the coupling factor, what is the new value of the total equivalent inductance?

Solution

$$\text{mutual inductance, } M = k \times \sqrt{L_1 L_2}$$
$$= 0.35 \times \sqrt{65 \times 85}$$
$$= 26 \text{ mH}$$

$$\text{parallel-aiding connection, } L_T^+ = \frac{L_1 L_2 - M^2}{L_1 + L_2 - 2M}$$
$$= \frac{65 \times 85 - 26^2}{65 + 85 - 2 \times 26}$$
$$= 49.5 \text{ mH}$$

$$\text{parallel-opposing connection, } L_T^- = \frac{L_1 L_2 - M^2}{L_1 + L_2 + 2M}$$
$$= \frac{65 \times 85 - 26^2}{65 + 85 + 2 \times 26}$$
$$= 24.0 \text{ mH}$$

POWER TRANSFORMERS

A power transformer consists of primary and secondary coils whose numbers of turns are N_p and N_s, Fig. 15-5A. These are wound on a common soft-iron core that reduces the leakage flux to a low value. In the case of the ideal transformer, the leakage flux is zero so that the mutual inductance, $M = \sqrt{L_p L_s}$, and the coupling factor, k, is unity. When an alternating current, I (rms), flows in the primary coil, it creates a magnetic flux that links with the secondary coil and induces the secondary voltage, E_s. With zero flux leakage, there are the same volts per turn associated with both the primary and secondary coils, and therefore

$$\text{turns ratio} = \frac{N_p}{N_s} = \frac{E_p}{E_s}$$

If N_s is greater than N_p, E_s is greater than E_p, and the transformer is referred to as a *step-up*. Likewise, if N_s is less than N_p, E_s is less than E_p, and the transformer is a *step-down*. The terms "step-up" and "step-down" normally

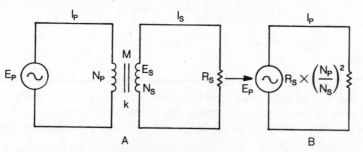

Fig. 15-5 Power transformer and its equivalent circuit.

refer to the voltage and not to the current. Note that if the primary and secondary coils are wound in the same sense, E_p and E_s are 180° out of phase.

The ideal transformer has zero power losses, and therefore the power input to the primary circuit equals the power output from the secondary circuit. Therefore,

$$E_pI_p = E_sI_s$$

and

$$\frac{E_p}{E_s} = \frac{I_s}{I_p} = \frac{N_p}{N_s}$$

This means that a step-up of the voltage level from the primary to the secondary circuit is accompanied by a corresponding reduction in the current level. The above equations can be rearranged as

$$E_p = \frac{E_sI_s}{I_p} = \frac{E_sN_p}{N_s}, I_p = \frac{E_sI_s}{E_p} = \frac{I_sN_s}{N_p}$$

$$E_s = \frac{E_pI_p}{I_s} = \frac{E_pN_s}{N_p}, I_s = \frac{E_pI_p}{E_s} = \frac{I_pN_p}{N_s}$$

$$N_p = \frac{E_pN_s}{E_s} = \frac{I_sN_s}{I_p}, N_s = \frac{E_sN_p}{E_p} = \frac{I_pN_p}{I_s}$$

TRANSFORMER EFFICIENCY

The practical power transformer has the following losses.

1. The copper loss, which is the power dissipated in the resistances of the primary and secondary windings.
2. The iron loss, which is dissipated in the core. This can be subdivided into (a) the eddy-current loss, which is caused by the flux cutting the soft iron and that can be reduced by laminating the core, and (b) the hysteresis loss, which is the result of rapidly magnetizing, demagnetizing, and remagnetizing the core during the cycle of the primary current.

These losses are taken into account by the efficiency percentage factor, η (Eta), which is defined by

$$\eta = \frac{\text{power output from the secondary circuit}}{\text{power input to the primary circuit}} \times 100\%$$

$$= \frac{E_sI_s}{E_pI_p} \times 100\%$$

or the secondary power output is

$$E_sI_s = E_pI_p \times \eta/100$$

Then

$$E_s = \frac{E_p I_p \eta}{I_s \times 100}, \; I_s = \frac{E_p I_p \eta}{E_s \times 100}$$

and

$$E_p = \frac{E_s I_s \times 100}{\eta I_s}, \; I_p = \frac{E_s I_s \times 100}{\eta E_p}$$

For the power losses in the transformer,

$$P_{loss} = \text{primary power} - \text{secondary power}$$

$$= \frac{100 - \eta}{100} \times \text{primary power}$$

$$= \frac{100 - \eta}{\eta} \times \text{secondary power}$$

For power transformers, the value of η normally exceeds 90%.

Note that if a steady direct current flows in the primary of a power transformer, the linkage flux is constant in magnitude and direction, and the voltage induced in the secondary coil is zero. However, if the steady dc voltage applied to the primary is mechanically chopped to produce a square wave, the transformer responds to such an input and an alternating voltage (not, however, a simple sine wave) is induced in the secondary. This secondary voltage can be rectified to produce a final dc output voltage which is larger than that of the dc input voltage.

REFLECTED RESISTANCE

If the secondary is loaded with a resistance R_s, then $R_s = E_s/I_s$. Because

$$E_p = \frac{E_s N_p}{N_s} \quad \text{and} \quad I_p = \frac{I_s N_s}{N_p}$$

Then

$$\frac{E_p}{I_p} = \frac{E_s N_p / N_s}{I_s N_s / N_p} = \frac{E_s}{I_s} \times \left(\frac{N_p}{N_s}\right)^2 = R_s \times \left(\frac{N_p}{N_s}\right)^2$$

The equation

$$\frac{E_p}{I_p} = R_s \times \left(\frac{N_p}{N_s}\right)^2$$

can be represented by the equivalent circuit shown in Fig. 15-5B. The expression $R_s \times (N_p/N_s)^2$ is the effective resistive load presented to the primary source and is referred to as the value of resistance reflected from the secondary circuit into the primary circuit due to the introduction of the secondary load R_s. If the turns ratio of the transformer is chosen so that the value of the reflected resistance, $R_s \times (N_p/N_s)^2$, is equal to the internal resistance of the primary source, the secondary load is then matched to the primary source for maximum power transfer to the secondary load.

If R_p is the resistance associated with the primary source, the condition for matching is

$$R_p = R_s\left(\frac{N_p}{N_s}\right)^2$$

and

$$R_s = R_p\left(\frac{N_s}{N_p}\right)^2$$

or

$$\frac{R_p}{R_s} = \left(\frac{N_p}{N_s}\right)^2$$

and

$$\frac{N_p}{N_s} = \sqrt{\frac{R_p}{R_s}}$$

Example 15-4

In Fig. 15-5A, $E_p = 110$ V, 60 Hz, $N_p = 1500$ turns, $N_s = 6000$ turns, and $R_s = 180$ Ω. Assuming a coupling factor of unity and that the transformer is 100% efficient, calculate the values of E_s, I_s, I_p, primary power, secondary power, and the reflected resistance.

Solution

$$\text{turns ratio} = \frac{N_p}{N_s} = \frac{1500}{6000} = 1:4$$

$$\text{secondary voltage, } E_s = E_p \times \frac{N_s}{N_p} = 110 \text{ V} \times 4$$
$$= 440 \text{ V}$$

$$\text{secondary current, } E_s = \frac{E_s}{R_s} = \frac{440 \text{ V}}{180 \text{ Ω}} = 2.444 \text{ A}$$

$$\text{primary current, } I_p = I_s \times \frac{N_s}{N_p} = 2.444 \times 4 = 9.776 \text{ A}$$

$$\text{primary power} = \text{secondary power} = E_p \times I_p = 110 \text{ V} \times 9.776$$
$$= 1075 \text{ W}$$

Check:

$$\text{secondary power} = E_s \times I_s = 440 \text{ V} \times 2.444 \text{ A}$$
$$= 1075 \text{ W}$$

$$\text{reflected resistance} = \frac{E_p}{I_p} = \frac{110 \text{ V}}{9.776 \text{ A}} = 11.25 \; \Omega$$

Check:

$$\text{reflected resistance} = R_s \times \left(\frac{N_p}{N_s}\right)^2 = \frac{180}{16}$$

$$= 11.25 \; \Omega$$

Example 15-5

In Fig. 15-5A, $E_p = 220$ V, 60 Hz, $I_p = 1.2$ A, $E_s = 55$ V, $R_s = 12 \; \Omega$. Calculate the values of the primary power, secondary power, and transformer efficiency.

Solution

$$\text{primary power} = E_p \times I_p = 220 \text{ V} \times 1.2 \text{ A} = 264 \text{ W}$$

$$\text{secondary power} = \frac{E_s^2}{R_s} = \frac{(55 \text{ V})^2}{12 \; \Omega} = 252 \text{ W}$$

$$\text{transformer power loss} = 264 - 252 = 12 \text{ W}$$

$$\text{transformer efficiency} = \frac{252}{264} \times 100 = 95\%$$

CHAPTER SUMMARY

☐ Mutually Coupled Circuits

$$\text{coupling factor, } k = \frac{M}{\sqrt{L_1 L_2}}$$

$$\text{mutual inductance, } M = k \times \sqrt{L_1 L_2}$$

☐ Mutually Coupled Coils

$$\text{series-aiding, } L_T^+ = L_1 + L_2 + 2M$$

$$\text{series-opposing, } L_T^- = L_1 + L_2 - 2M$$

$$\text{parallel-aiding, } L_T^+ = \frac{L_1 L_2 - M^2}{L_1 + L_2 - 2M}$$

$$\text{parallel-opposing, } L_T^- = \frac{L_1 L_2 - M^2}{L_1 + L_2 + 2M}$$

☐ Power Transformer
100% efficiency, coupling factor = 1.

$$\text{turns ratio} \quad \frac{N_p}{N_s} = \frac{E_p}{E_s} = \frac{I_s}{I_p}$$

$$\text{primary power, } E_p I_p = \text{secondary power, } E_s I_s$$

Resistance reflected into the primary circuit equals

$$R_s \times \left(\frac{N_p}{N_s}\right)^2$$

☐ Practical Transformer
Coupling factor less than 1.

$$\text{transformer efficiency, } \eta = \frac{\text{secondary power}}{\text{primary power}} \times 100\%$$

$$= \frac{E_s I_s}{E_p I_p} \times 100\%$$

16
AC Circuit Analysis

IN THIS CHAPTER on ac circuit analysis, you will learn:

☐ How to use the j operator in the solution of complicated ac circuits.
☐ The use of Kirchhoff's Laws in solving ac circuits with more than one source.
☐ How to use mesh currents in the solution of ac circuits involving a number of loops.
☐ How to apply the superposition theorem in solving an ac circuit with more than one source.
☐ About the use of nodal analysis in solving an ac circuit in which there are a number of current sources and it is necessary to find the voltage at a particular node (with reference to ground).
☐ How to use Millman's Theorem in combining a number of constant-current alternating sources that are connected directly in parallel.
☐ How to Thévenize part of an ac circuit and then determine the (load) current associated with the remainder of the circuit.
☐ How to Nortonize part of an ac circuit and then determine the (load) current associated with the remainder of the circuit.

In Chapter 8, we explored the various network theorems and their application to dc circuit analysis. If the same theorems are to be used in sinewave ac circuit analysis we must recognize the added difficulties of reactance, impedance, and phase difference, all of which do not exist in dc circuits. Fortunately, these differences can be resolved by the use of the j operator and the rules of complex algebra that we discussed in Chapter 13. We are then able to solve network problems without any reference to phasor diagrams. In such problems, all the voltages and currents are assumed to have the same frequency.

In networks that involve more than one voltage (or current) source, these sources are not necessarily in phase. Each source (for example, $8\underline{/75°}$ V or $3\underline{/-40°}$ A) therefore has its own polar angle that is related to some reference sine wave.

KIRCHHOFF'S LAWS

I will now restate Kirchhoff's Laws as they apply to ac circuits.

Kirchhoff's Voltage Law (KVL)

The phasor sum of the source voltages and the voltage drops around any closed electrical loop is always zero.

Kirchhoff's Current Law (KCL)

The phasor sum of the currents existing at any electrical junction point is zero. In dc circuits, we adopted a convention in order to distinguish between positive and negative dc voltages as they appeared in the algebraic equations. However, we cannot assign a constant direction to an alternating current or a fixed polarity to an alternating voltage. Instead we will give an instantaneous polarity ("+" or "−") to a source voltage and then indicate the associated direction of the (electron) current. Figure 16-1 illustrates these points.

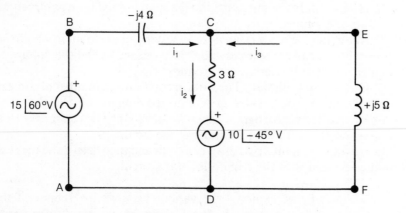

Fig. 16-1 Circuit to illustrate the application of Kirchhoff's Laws.

Example 16-1

In the circuit of Fig. 16-1, find the value of the current i_3.

Solution

The KVL equation for the loop ABCDA is

$$i_1 \times (-j4) - i_2 \times (3) = 15\underline{/60°} - 10\underline{/-45°}$$
$$= 7.5 + 12.99 - 7.07 + j7.07$$
$$= 0.43 + j20.06$$

Note that the two voltage sources are opposing, in accordance with their polarities as shown.

The KVL equation for the loop CEFDC is

$$i_2 \times 3 + i_3 \times j5 = 10\underline{/-45°} = 7.07 - j7.07$$

At the junction point, C, the KCL equation is

$$i_3 = i_1 + i_2$$

or

$$i_2 = i_3 - i_1$$

The two preceeding equations yield

$$3i_3 - 3i_1 + i_3 \times j5 = 7.07 - j7.07$$
$$- 3i_1 + i_3(3 + j5) = 7.07 - j7.07 \quad \text{(Eq. 16-1)}$$

From the KVL equation for the loop ABCDA

$$i_1 \times (-j4) - (i_3 - i_1) \times 3 = 0.43 + j20.06$$
$$i_1 \times (3 - j4) - 3i_3 = 0.43 + j20.06$$

Adding the last two equations gives

$$i_1 \times (-j4) + i_3 \times (j5) = 7.5 + j12.99 \quad \text{(Eq. 16-2)}$$

Multiplying Eq. 16-1 by $-j4$ and Eq. 16-2 by -3,

$$i_1 \times j12 + i_3(-j12 + 20) = -j28.28 - 28.28 \quad \text{(Eq. 16-3)}$$
$$i_1 \times j12 - i_3 \times j15 = -22.5 - j38.97 \quad \text{(Eq. 16-4)}$$

Subtracting Eq. 16-4 from Eq. 16-3,

$$i_3(-j12 + 20 + j15) = -5.78 + j10.69$$
$$i_3 = \frac{-5.78 + j10.69}{20 + j3}$$

$$\text{current, } i_3 = \frac{12.15\underline{/118.4°}}{20.22\underline{/8.53°}} = 0.60\underline{/109.9°} \text{ A}$$

MESH-CURRENT ANALYSIS

In Chapter 8, we explored the use of mesh-current analysis as applied to dc circuits. The same principles apply to ac circuits so that we can replace the three branch currents of Fig. 16-1 with the two mesh currents i_1 and i_2 (Fig. 16-2). We will then be involved only with two simultaneous equations rather than the three equations that appeared in the Kirchhoff analysis. Moreover, you can write the mesh equations down simply by inspection.

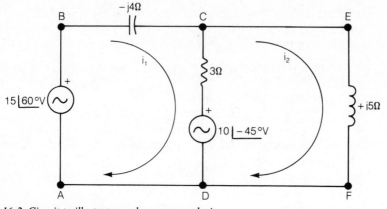

Fig. 16-2 Circuit to illustrate mesh current analysis.

Example 16-2

In Fig. 16-2, use the method of mesh-current analysis to determine the value of i_2.

Solution

As far as the polarities of the voltage sources are concerned, we observe the normal KVL convention.

In the mesh ABCDA,

$$i_1(3 - j4) - i_2 \times 3 = -15\underline{/60°} + 10\underline{/45°}$$

$$i_1(3 - j4) - i_2 \times 3 = -7.5 - j12.99 + 7.07 - j7.07 = 0.43 - j20.06$$

In the mesh CEFDC,

$$-3\,i_1 + i_2 \times (3 + j5) = -10\underline{/45°} = -7.07 + j7.07$$

Multiply the first equation by 3 and the second equation by $3 - j4$. This yields

$$i_1(9 - j12) - i_2 \times 9 = -1.29 - j60.18$$
$$-i_1(9 - j12) + i_2(3 + j5)(3 - j4) = (-7.07 + j7.07)(3 - j4)$$

or

$$-i_1(9 - j12) + i_2(29 + j3) = 7.07 + j49.49$$

Adding the two preceding equations yields

$$i_2(20 + j3) = 5.78 - j10.69$$

Therefore,

$$\text{current, } i_2 = \frac{5.78 - j10.69}{20 + j3} = \frac{12.15\underline{/-61.6°}}{20.22\underline{/8.35°}} = 0.60\underline{/-69.1°} \text{ A}$$

Note that the mesh current, i_2, is 180° out of phase with the Kirchhoff branch current, i_3, in Example 16-1. This is due to our assuming opposite directions for these two currents.

SUPERPOSITION THEOREM

This theorem was applied to dc currents in Chapter 8. For ac circuits, the theorem can be restated as follows:

> "If a network of linear impedances contains more than one alternating source, the current flowing at any point is the phasor sum of the currents that would flow at that point if each alternating source were considered separately, with all other alternating sources replaced by impedances equal to their internal impedances. This would involve replacing each voltage source by a short circuit and each current source by an open circuit."

The following example illustrates the application of the superposition theorem.

Example 16-3

In the circuit of Fig. 16-3, use the superposition theorem to determine the value of the current i_L.

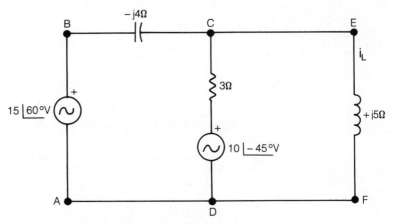

Fig. 16-3 Circuit to illustrate the superposition theorem.

Note: this is the same circuit as appeared in Figs. 16-1 and 16-2.

Solution

Step 1. Replace the $10\underline{/-45°}$ V source by a short circuit (Fig. 16-4A). The total impedance, z_{1T}, then presented to the $15\underline{/-60°}$ V source is

$$z_{1T} = -j4 + \frac{3 \times j5}{3 + j5} = -j4 + \frac{j15(3 - j5)}{3^2 + 5^2}$$

$$= -j4 + \frac{75 + j45}{34}$$
$$= -j4 + 2.21 + j1.32$$
$$= 2.21 - j2.68 \ \Omega$$

Then

$$i_{1T} = \frac{15\underline{/60°} \ V}{2.21 - j2.28 \ \Omega}$$

and

$$i_{1L} = \frac{15\underline{/60°}}{2.21 - j2.68} \times \frac{3}{3 + j5}$$
$$= \frac{45\underline{/60°}}{3.47\underline{/-50°} \times 5.83\underline{/59.04°}}$$
$$= 2.22\underline{/51.46°} \ A$$

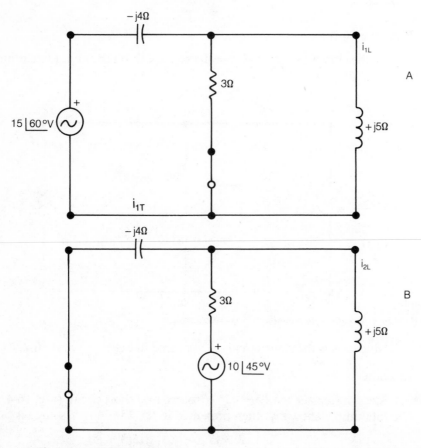

Fig. 16-4 *The superposition principle.*

Step 2. Replace the $15\underline{/60°}$ V source by a short circuit (Fig. 16-4B). The total impedance, z_{2T}, presented to the $10\underline{/-45°}$ V source is

$$z_{2T} = 3 + \frac{(-j4) \times j5}{j5 - j4} = 3 + \frac{20}{j1} = 3 - j20 \ \Omega$$

Then

$$i_{2T} = \frac{10\underline{/-45°} \ V}{3 - j20 \ \Omega}$$

and

$$
\begin{aligned}
i_{2L} &= \frac{10\underline{/-45°}}{3 - j20} \times \frac{(-j4)}{j4 + j5} = \frac{-40\underline{/-45°}}{3 - j20} \\
&= \frac{-40\underline{/-45°}}{20.22\underline{/-81.47°}} \\
&= -1.98\underline{/36.47°} \ A
\end{aligned}
$$

Using the principle of superposition,

$$
\begin{aligned}
\text{current, } i_L = i_{L1} + i_{L2} &= 2.22\underline{/51.46°} - 1.98\underline{/36.47°} \\
&= 1.38 + j1.736 - 1.592 - j1.177 \\
&= -0.212 + j0.559 \\
&= 0.6\underline{/110°} \ A
\end{aligned}
$$

This is the same value as was obtained for the Kirchhoff branch current, i_3, in Example 16-1.

NODAL ANALYSIS

In Chapter 8 we studied the application of nodal analysis to dc circuits. For ac circuits, we will have to modify the analysis in order to take into account the various phase relationships between the currents entering and leaving a node. The statement for nodal analysis is as follows:

> "At a node the algebraic phasor sum of the source currents entering = the phasor sum of the currents leaving through the impedances."

As an example, we will redraw the circuit of Fig. 16-1 so that it appears as shown in Fig. 16-5A. The point N is then a node.

Example 16-4

In Fig. 16-5, determine the voltage at the point N.

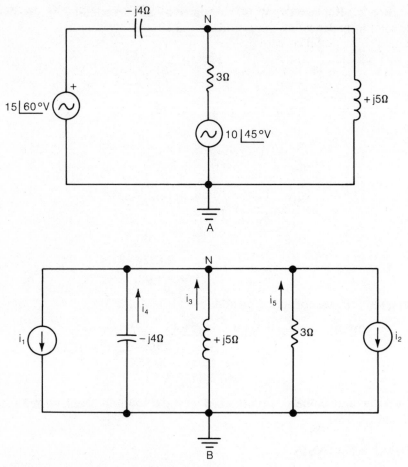

Fig. 16-5 Circuit to illustrate nodal analysis.

Solution

Regard the capacitive reactance of $-j6$ Ω as the internal impedance of the $15\underline{/60°}$ V source. Similarly, the resistance of 3 Ω is the internal impedance of the $10\underline{/-45°}$ V source. Convert both voltage sources into their equivalent current generators (Fig. 16-5B).
Then

$$i_1 = \frac{15\underline{/60°}\text{ V}}{-j4\ \Omega} = \frac{15\underline{/60°}\text{ V}}{4\underline{/-90°}\ \Omega} = 3.75\underline{/150°}\text{ A}$$

and

$$i_2 = \frac{10\underline{/-45°}\text{ V}}{3\ \Omega} = 3.333\underline{/-45°}\text{ A}$$

The currents i_1 and i_2 are leaving the node, while the currents i_3, i_4, and i_5 are entering the point N.

Therefore the nodal equation is

$$3.75\underline{/150°} + 3.333\underline{/-45°} = \frac{v_N}{j5} + \frac{v_N}{-j4} + \frac{v_N}{3}$$

where v_N is the alternating voltage at the node.

Then

$$-3.248 + j1.875 + 2.352 - j2.357 = v_N(-j0.2 + j0.25 + 0.333)$$
$$v_N(0.333 + j0.05) = -0.891 - j\,0.482$$
$$v_N = \frac{1.013\underline{/-151.6°}}{0.337\underline{/8.5°}} \approx 3\underline{/-160°} \text{ V}$$

Check:

current through the inductor is

$$i_3 = \frac{v_N}{j5}$$
$$= \frac{3\underline{/-160°} \text{ V}}{5\underline{/90°} \text{ } \Omega}$$
$$= 0.6\underline{/-250°}$$
$$= 0.6\underline{/110°} \text{ A}$$

This value agrees with the results obtained with the previous methods of analysis.

MILLMAN'S THEOREM

Millman's Theorem for the solution of ac circuits can be stated as follows:

> "Any number of constant alternating current sources that are connected *directly* in parallel can be combined into a single alternating-current source whose total current is the phasor sum of the individual source currents and whose total internal impedance is the result of combining the individual source impedances in parallel."

Example 16-5

In the circuit of Fig. 16-5B, use Millman's Theorem to determine the value of the current, i_3.

Solution

The currents, i_1 and i_2, flow in the same direction. Consequently, the value, i_T, of the total equivalent current generator, is

$$i_T = i_1 + i_2 = 3.75\underline{/150°} + 3.333\underline{/-45°}$$
$$= -0.891 - j\,0.482 \text{ A.}$$

The total impedance, z_T, of $-j4\ \Omega \mathbin{/\!/} j5\ \Omega \mathbin{/\!/} 3\ \Omega$ is given by

$$\frac{1}{z_T} = \frac{1}{-j4} + \frac{1}{j5} + \frac{1}{3} = +j0.25 - j0.2 + 0.333$$
$$= 0.333 + j0.05$$

Using the current-division rule (CDR),

$$\text{current, } i_3 = i_T \times \frac{z_T}{j5} = \frac{-0.891 - j0.482}{j5(0.333 + j0.05)}$$
$$= \frac{1.013\underline{/-151.6°}}{-0.25 + j1.667}$$
$$= \frac{1.013\underline{/-151.6°}}{1.686\underline{/98.5°}} \approx 0.6\underline{/110°} \text{ A}$$

This is the same result that we achieved with all the other analytical methods.

THÉVENIN'S THEOREM

The procedure in Théveninizing an ac circuit is the same as outlined in Chapter 8. In the case of ac circuits, the statement of the theorem is as follows:

> "The current in a load impedance connected between two terminals X and Y of a network of impedances and alternating sources is the same as if that load impedance were connected to a single alternating source whose output is the open-circuit voltage as measured between X and Y and whose internal impedance is the impedance of the network looking back into the terminals X and Y with all alternating sources replaced by impedances equal to their internal impedances."

We will now use Thévenin's theorem to solve the circuit of Fig. 16-6A.

Example 16-6

In Fig. 16-6A, derive the Thévenin equivalent circuit between terminals X and Y, and then obtain the load current in its polar form.

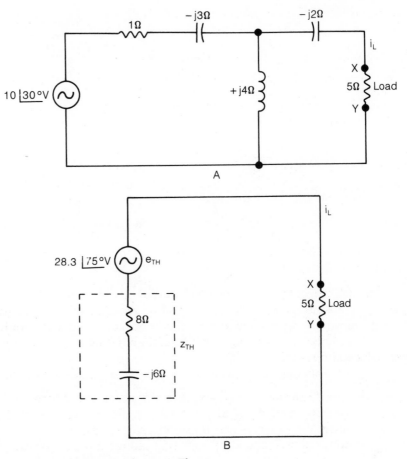

Fig. 16-6 Circuit to illustrate Thévenin's Theorem.

Solution

Step 1. Remove the load and calculate the open-circuit voltage between X and Y. Using the voltage division rule,

$$\text{Thévenin voltage, } e_{TH} = 10\underline{/30°} \times \frac{j4}{(1-j3)+j4} = 10\underline{/30°} \times \frac{4\underline{/90°}}{1+j1}$$

$$= \frac{10\underline{/30°} \times 4\underline{/90°}}{2\underline{/45°}}$$

$$= 28.3\underline{/75°} \text{ V}$$

Step 2. Replace the voltage source by a short circuit and calculate the value of the impedance phasor between X and Y.

$$\text{Thévenin impedance, } z_{TH} = -j2 + \frac{j4 \times (1-j3)}{j4+(1-j3)} = -j2 + \frac{j4+12}{1+j1}$$

$$= -j2 + \frac{(j4 + 12)(1 - j1)}{2}$$
$$= -j2 + 8 - j4$$
$$= 8 - j6 \ \Omega$$

Step 3   Replace the load in the Thévenin equivalent circuit of Fig. 16-6B. The load current is then

$$i_L = \frac{e_{TH}}{z_{TH} + R_L} = \frac{28.3\underline{/75°}}{8 - j6 + 5} = \frac{28.3\underline{/75°}}{13 - j6}$$

$$= \frac{28.3\underline{/75°}}{14.3\underline{/-24.78°}}$$
$$= 1.98\underline{/99.78°} \text{ A}$$

NORTON'S THEOREM

Norton's Theorem was applied to dc circuits in chapter 8. For ac circuits there is no difference in the procedure required in Nortonizing a complex network of alternating sources and impedances so that the equivalent Norton alternating source again will be of the constant-current type with its associated impedance in parallel.

The formal statement of Norton's theorem for ac circuits is as follows:

> "The alternating current in a load connected between two output terminals, X and Y, of a complex network containing alternating sources and impedances, is the same as if the load were connected to a constant alternating source whose current, i_N, is equal to the short-circuit current measured between X and Y. This constant alternating current source, i_N, is placed in parallel with an impedance, z_N, which is equal to the impedance of the network looking back into the terminals X and Y with all sources replaced by impedances equal in value to their internal impedances."

The last part of this statement involves substituting short circuits for all constant alternating-voltage sources and open circuits for all constant alternating-current sources.

To compare Norton's Theorem with Thévenin's Theorem we will again determine the value of the load current in Fig. 16-6A.

Example 16-7

In Fig. 16-6A, derive the Norton equivalent circuit between the terminals, X and Y, and then obtain the load current in its polar form.

Solution

Step 1. Remove the load and replace it by a short circuit. The total impedance then presented to the voltage source is

$$z_T = 1 - j3 + \frac{j4 \times (-j2)}{j4 + (-j2)}$$

$$= 1 - j3 + \frac{8}{j2} = 1 - j3 - j4$$

$$= 1 - j7 \; \Omega$$

The current drawn from the source is

$$i_T = \frac{10\underline{/30°}}{1 - j7} \; A$$

and the short circuit current between the terminals, X, Y is

$$\text{norton current, } i_N = \frac{10\underline{/30°}}{1 - j7} \times \frac{j4}{j4 + (-j2)}$$

$$= \frac{20\underline{/30°}}{7.07\underline{/-81.87°}} = 2.83\underline{/111.87°} \; A$$

Step 2. The Norton impedance, z_N, is the same as the Thévenin impedance, z_{TH}, and therefore $z_N = 8 - j6 \; \Omega$ (Fig. 16-6B).

Step 3. Remove the short circuit and replace the load in the Norton equivalent circuit of Fig. 16-7. By the current division rule,

$$\text{load current, } i_L = i_N \times \frac{8 - j6}{8 - j6 + 5} = 2.83\underline{/111.87°} \times \frac{8 - j6}{13 - j6}$$

$$= \frac{2.83\underline{/111.87°} \times 10\underline{/-36.87°}}{14.32\underline{/-24.78°}}$$

$$= 1.98\underline{/99.78°} \; A$$

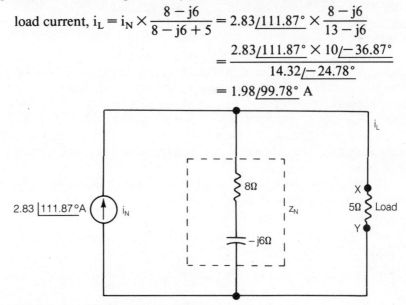

Fig. 16-7 Circuit to illustrate Norton's Theorem.

CHAPTER SUMMARY

☐ Kirchhoff's Voltage Law (KVL)

The phasor sum of the alternating voltage sources and the voltage drops around any closed electrical loop is zero.

☐ Kirchhoff's Current Law (KCL)

The phasor sum of the currents existing at any junction point is zero.

☐ Superposition Theorem

In a network of linear impedances containing more than one alternating source, the current flowing at any point is the phasor sum of the currents that would flow at that point if each alternating source is considered separately with all other sources replaced by their equivalent internal impedances.

☐ Nodal Analysis

The phasor sum of the alternating source currents entering the node is equal to the phasor sum of the currents leaving through the impedances.

☐ Millman's Theorem

Any number of alternating-current sources that are connected directly in parallel can be converted into a single alternating-current source whose total alternator current is the phasor sum of the individual source currents and whose internal impedance is the result of combining the individual source impedances in parallel.

☐ Thévenin's Theorem

The alternating current in the impedance load connected between two terminals X and Y of a complex network of impedances and alternating sources is the same as if that load were connected across a simple alternator whose voltage, e_{TH}, is the *open-circuit* value measured between X and Y and whose *series* internal impedance, z_{TH}, is the impedance of the network looking back into the terminals, X and Y, with all sources replaced by impedances equal to their internal impedances.

☐ Norton's Theorem

The alternating current in a load connected between two output terminals X and Y of a complex network containing alternating sources and impedances is the same as if that load were connected to an alternating source whose alternator current, i_N, is equal to the *short circuit* value measured between X and Y. This current alternator is placed in *parallel* with an impedance, z_N, that is equal to the network's impedance looking back into the terminals X, Y with all sources replaced by impedances equal to their internal impedances.

The relationships between Thévenin and Norton alternators are

$$e_{Th} = i_N \times z_N, \, i_N = \frac{e_{Th}}{z_{Th}}$$

$$z_{Th} = z_N$$

17

AC Generators:
Polyphase Systems

As we investigate three-phase operation, you will learn:

☐ The principles of the single-phase revolving-field alternator.
☐ The advantages of two-phase and three-phase operation compared to single phase.
☐ The reasons for using wye and delta connections with three-phase operation.
☐ The relationship between the line and phase voltages of a wye connection to a balanced system of loads.
☐ The relation between the line and phase currents of a delta connection.
☐ How to calculate the total power output of a three-phase alternator.

In Chapter 9, the basic alternator contained a revolving armature whose conductors cut the flux of a stationary magnetic field. The alternating voltage generated then was taken through slip-rings to the load. The contacts between these slip-rings and the carbon brushes were subject to friction wear and sparking; moreover, they were liable to arc over at high voltages. These problems are overcome in the revolving field alternator of Fig. 17-1.

THE PRACTICAL ALTERNATOR

In this type of generator, a dc source drives a direct current through sliprings, brushes, and the windings on a rotor that is driven around by mechanical means. This creates a rotating magnetic field that cuts the conductors embedded in the surrounding stator. An alternating voltage then appears between the ends, S,F, of the stator winding. Because S and F are fixed terminals, there are no sliding contacts, and the whole of the stator winding can be insulated continuously.

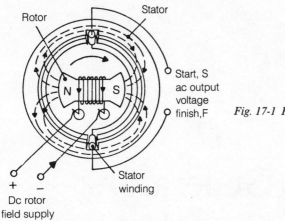

Fig. 17-1 Revolving-field alternator.

The alternator of Fig. 17-1 has two poles so that one complete rotation of the rotor generates one cycle of ac in the stator. Therefore, in order to generate 60 Hz, the rotor must turn at 60 revolutions per second, or 60 × 60 = 3600 rpm. However, in the four pole machine of Fig. 17-2, one revolution produces two cycles of ac voltage, and therefore it only requires a rotor speed of 1800 rpm to generate a 60-Hz output. It follows that

$$\text{generated frequency, } f = \frac{Np}{60} \text{ Hz}$$

where

$$N = \text{rotor speed in rpm}$$
$$p = \text{number of pairs of poles}$$

The power delivered to a resistive load by a single-phase alternator is fluctuating at twice the line frequency (Chapter 9). This presents a problem because the load on the mechanical source of energy, and therefore the necessary torque, is not constant.

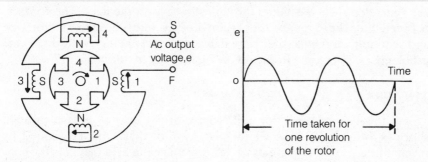

Fig. 17-2 Alternator with four-pole rotor.

Example 17-1

An eight-pole rotor is revolving at 750 rpm. What is the value of the generated frequency?

Solution

$$\text{generated frequency, } f = \frac{Np}{60} = \frac{750 \times 4}{60} = 50 \text{ Hz}$$

THE TWO-PHASE ALTERNATOR

In this type of alternator, two equal coils are mounted on the stator with their axes separated by 90°. Fig. 17-3A shows a simplified arrangement of the windings which are cut by the magnetic flux of a two-pole rotor. The EMFs, e_1 and e_2, induced in the windings will therefore be equal in magnitude but 90° out of phase (Fig. 17-3B). These voltages may therefore be represented by

$$e_1 = E_{max} \sin \omega t$$

and

$$e_2 = E_{max} \sin (\omega t + \pi/2) = E_{max} \cos \omega t$$

where E_{max} is the peak value of the phase voltage. The fact that e_2 leads e_1 by 90° is a result of the assumed clockwise movement of the rotor. This phase relationship commonly is represented by drawing the two stator windings 90° apart (Fig. 17-3C). Each of the windings can be connected to separate loads, but it is more convenient to use a neutral line so that the number of lines required is reduced from four to three as in Fig. 17-3D. Assuming identical (balanced) resistive loads, the instantaneous powers are

$$p_1 = \frac{e_1^2}{R} = \frac{E_{max}^2 \sin^2 \omega t}{R}$$

and

$$p_2 = \frac{e_2^2}{R} = \frac{E_{max}^2 \cos^2 \omega t}{R}$$

Then

$$p_1 + p_2 = \frac{E_{max}^2}{R}$$

which is independent of time. The total instantaneous power of the two-phase alternator is therefore constant as opposed to the fluctuating power output of the single-phase machine. Note that this advantage is achieved only if the loads are balanced.

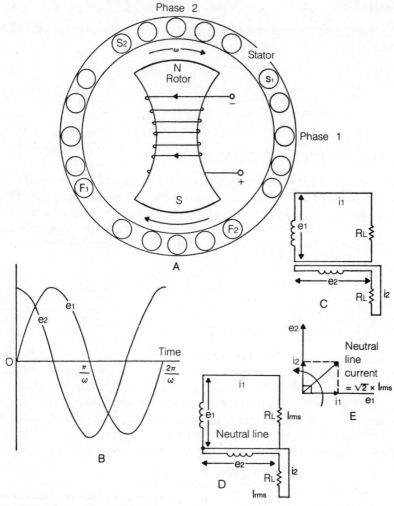

Fig. 17-3 The two-phase alternator.

If the current in each load is I_{rms}, the neutral line current is

$$\sqrt{I_{rms}^2 + I_{rms}^2} = \sqrt{2}\, I_{rms}$$

(Fig. 17-3E).

The total current in the three lines is therefore $(2 + \sqrt{2})I_{rms}$. If the same loads are connected in parallel across a single-phase alternator, the total current in the two supply lines is $4\, I_{rms}$. A single-phase system therefore requires more copper to supply a particular load at a given voltage.

If the two voltages are connected to two pairs of coils whose axes are perpendicular, the two fluxes associated with the coils combine to produce a rotating magnetic field whose angular frequency is equal to that of the alternating voltage. This principle is used in ac induction and synchronous motors.

Compared to single phase alternators, high power two and three phase generators require smaller stators, are more efficient and are less subject to vibration.

Example 17-2

A two-phase alternator has a four-pole rotor whose speed is 1600 rpm. What is the value of the generated frequency? The rms voltage of each phase is 120 V, and the resistive loads are balanced so that each has an effective resistance of 40 Ω. Find the current in the neutral line and the total instantaneous power output of the generator.

Solution

The generated frequency is independent of the number of phases.

$$\text{generated frequency} = \frac{1500 \times 2}{60} = 50 \text{ Hz}$$

$$\text{load current for each phase} = \frac{120 \text{ V}}{40 \text{ }\Omega} = 3 \text{ A rms}$$

$$\text{neutral line current} = \sqrt{3^2 + 3^2} = 4.242 \text{ A rms}$$

$$\text{total instantaneous power} = \frac{E_{max}^2}{R} = \frac{(120 \sqrt{2})^2}{40}$$

$$= 720 \text{ W}$$

This is also the average power delivered over the cycle.

Note: for the equivalent single-phase alternator, the total load is equal to 40 $\Omega/2 = 20$ Ω. The maximum instantaneous power is

$$\frac{(120 \sqrt{2})^2}{20} = 1440 \text{ W}$$

and the average power over the cycle is 1440 W/2 = 720 W. The total of the currents in the two supply lines is 2 $\times$ 120/20 = 12 A; this compares with the two phase alternator in which the total of the currents in the two supply lines and the neutral line is (2 $\times$ 3) + 4.242 = 10.242 A.

THE THREE-PHASE ALTERNATOR

As shown in Fig. 17-4A, the three-phase alternator has three single-phase windings that are so spaced on the stator that the voltage induced in each winding is 120° out of phase with the voltages in the other two windings (Fig. 17-4B). It should be emphasized that in the three-phase alternator the windings are independent of each other and could be connected to separate loads; this would require a six-line system (Fig. 17-4C).

The three voltage waveforms can be represented by their equivalent phasors as illustrated in Fig. 17-4D.

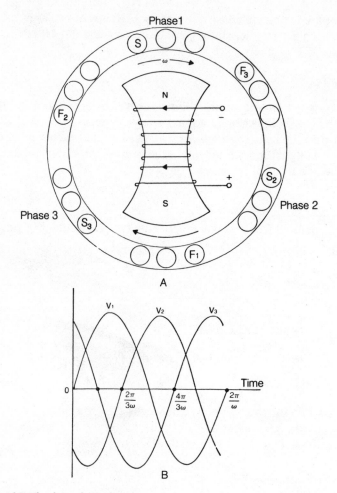

Fig. 17-4 A&B The three-phase alternator.

If v_1 is used as the reference phase voltage, and the rotor is assumed to be revolving in the clockwise direction, then

$$v_1 = E_{max} \sin \omega t$$

$$v_2 = E_{max} \sin \left(\omega t - \frac{2 \pi}{3} \right) = - \frac{E_{max} \sin \omega t}{2} - \frac{\sqrt{3}}{2} E_{max} \cos \omega t$$

$$v_3 = E_{max} \sin \left(\omega t - \frac{4 \pi}{3} \right) = E_{max} \sin \left(\omega t + \frac{2 \pi}{3} \right)$$

$$= - \frac{E_{max} \sin \omega t}{2} + \frac{\sqrt{3}}{2} E_{max} \cos \omega t$$

From these equations it is apparent that

$$v_1 + v_2 + v_3 = 0$$

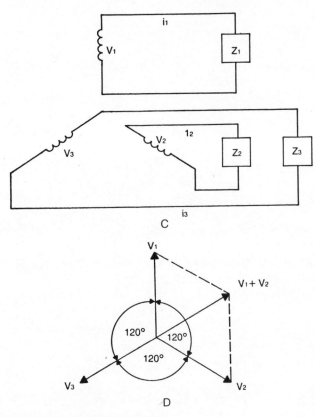

Fig. 17-4 C&D Six-wire three-phase connection.

This is also obvious from Fig. 17-4D, in which the phasor addition $v_1 + v_2$ is equal in magnitude to v_3 but is exactly opposite in direction. At any point in time, the sum of the instantaneous phase voltages is therefore zero. It is then possible to connect three ends of the windings to a common point and the other ends to separate terminals. A single neutral line can be connected to the common point to produce a four-wire wye (Y) system (chapter 8).

Assume that the three loads are connected directly across the phase windings, (Fig. 17-5A). The loads are balanced in the sense that they all have the same magnitude and power factor (phase angle), which in this case is assumed be lagging. The three load currents will also be equal in magnitude and 120° out of phase with each other (Fig. 17-5B) so that their phasor sum as carried by the neutral line is zero. The neutral line therefore can have a low current capacity because any current that it carries can exist only as a result of imbalance between the loads. Alternatively, the neutral line can be replaced by a ground return. The saving in the copper required by the supply lines therefore is greater for three-phase than for two-phase operation (see Example 17-3).

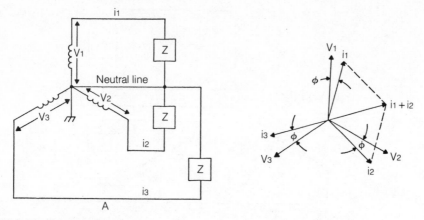

Fig. 17-5 The four-wire wye system.

If each of the balanced loads is equal to a resistance, R, the instantaneous powers delivered by the phases are

$$p_1 = \frac{E_{max}^2}{R} \sin^2 \omega t$$

$$p_2 = \frac{E_{max}^2}{R} \sin^2 \left(\omega t - \frac{2\,\pi}{3} \right)$$

$$= \frac{E_{max}^2}{R} \left(\frac{-\sin \omega t + \sqrt{3} \cos \omega t}{2} \right)^2$$

$$p_3 = \frac{E_{max}^2}{R} \sin^2 \left(\omega t + \frac{2\,\pi}{3} \right)$$

$$= \frac{E_{max}^2}{R} \left(\frac{\sin \omega t + \sqrt{3} \cos \omega t}{2} \right)^2$$

Then

$$p_1 + p_2 + p_3 = \frac{E_{max}^2}{R} \left[\begin{array}{c} \sin^2 \omega t + \dfrac{\sin^2 \omega t}{4} + \dfrac{\sin^2 \omega t}{4} \\[2mm] + \dfrac{3}{4} \cos^2 \omega t + \dfrac{3}{4} \cos^2 \omega t \end{array} \right]$$

$$= \frac{3}{2} \times \frac{E_{max}^2}{R} = \frac{3\,E_{rms}^2}{R}$$

The total instantaneous power for a balanced three system is therefore constant and independent of time.

Example 17-3

A three-phase alternator has a four-pole rotor whose speed is 1800 rpm. What is the value of the generated frequency? The rms value of each phase is 120 V, and the resistive loads are balanced so that each has an effective resistance of 60 Ω. Find the total of the line and neutral currents and the total instantaneous power output of the alternator.

Solution

$$\text{generated frequency} = \frac{1800 \times 2}{60} = 60 \text{ Hz}$$

$$\text{load current for each phase} = \frac{120 \text{ V}}{60} = 2 \text{ A}$$

Neutral line current is zero for a balanced load system.

$$\text{total of the load and neutral currents} = 3 \times 2 \text{ A} = 6 \text{ A}$$
$$\text{total instantaneous power} = \text{average power over the cycle}$$

$$= 3 \times \frac{(120 \text{ V})^2}{60} = 720 \text{ W}$$

Note: for the equivalent two-phase alternator, the load on each phase would be $2 \times 60/3 = 40 \ \Omega$. The supply line current for each phase is then 120 V/40 Ω = 3 A, and the neutral line current is $3 \times \sqrt{2} = 4.242$ A. The total of the line and neutral currents is $(2 \times 3) + 4.242 = 10.242$ A.

Total instantaneous power = average power over the cycle

$$= 2 \times \frac{(120 \text{ V})^2}{40 \ \Omega} = 720 \text{ W}$$

For the single-phase alternator, the total load would be $60/3 = 20 \ \Omega$. The total current in the two supply lines is then 2×120 V/20 $\Omega = 12$ A, and the average power over the cycle is

$$\frac{(120 \text{ V})^2}{20 \ \Omega} = 720 \text{ W}$$

THE WYE CONNECTION

As explained earlier, the EMF applied to the balanced loads was the voltage that existed between each of the lines and ground (or the neutral line). This EMF normally is referred to as the phase voltage, V_p. However, with three-phase systems it is normal to make use of the line voltage as well as the phase voltage. As shown in Fig. 17-6A, the line voltage, V_L, is the EMF existing between any two of the lines. One of the line voltages therefore exists between points X, Y and this voltage is related to the EMFs generated in the first and second phases of the stator windings.

In determining the relationship between V_L and V_p, it is important to realize that the correct stator terminals must be connected to the common point if the phasor voltages are to be represented as 120° apart; for example, if the two terminals of one winding are reversed, its phase voltage is shifted by 180°. The phase voltages v_X, v_Y, and v_Z are the ac voltages monitored at the points X, Y, Z with respect to the common point 0. Consequently, the line

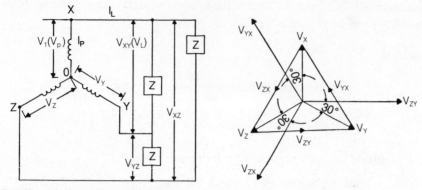

Fig. 17-6 The line voltage of the wye system.

voltage v_{XY} is the ac voltage at X with respect to Y and therefore is the voltage difference between the two points. In other words $v_{XY} = v_X - v_Y$. The phasor line voltages are

$$v_{XY} = v_X - v_Y$$
$$v_{YZ} = v_Y - v_Z$$
$$v_{ZX} = v_Z - v_X$$

If

$$v_X = E\underline{/0}, \; v_Y = E\underline{/-2\,\pi/3}, \; v_Z = E\underline{/+2\,\pi/3}$$
$$v_{XY} = E\underline{/0} - E\underline{/-2\,\pi/3}$$
$$= E - E\left[\left(-\frac{1}{2}\right) + j(-\sqrt{3}/2)\right]$$
$$= E(3/2 + j\,\sqrt{3}/2) = \sqrt{3}\,E\underline{/\pi/6}$$
$$v_{YZ} = E\underline{/-2\,\pi/3} - E\underline{/+2\,\pi/3}$$
$$= E\left[\left(-\frac{1}{2}\right) + j(-\sqrt{3}/2)\right] - E\left[\left(-\frac{1}{2}\right) + j(+\sqrt{3}/2)\right]$$
$$= \sqrt{3}\,E\underline{/-\pi/2}$$
$$v_{ZX} = E\underline{/+2\,\pi/3} - E\underline{/0}$$
$$= E\left[\left(-\frac{1}{2}\right) + j(+\sqrt{3}/2)\right] - E\underline{/0}$$
$$= E[(-3/2) + j(+\sqrt{3}/2)]$$
$$= \sqrt{3}\,E\underline{/5\,\pi/6}$$

These equations all show that $V_L = \sqrt{3}\,V_p$. Like the phase voltages, the line voltages are 120° apart and are shifted by 30° from the phase voltages (Fig. 17-6B).

When three balanced loads are connected between the lines, the three line currents are also 120° apart. Because each line is connected directly to a terminal of a phase winding, the line current, I_L, must be equal to the phase current, I_p.

Example 17-4

In a three phase system, one phase voltage is 120$\underline{/50°}$ volts. Balanced loads, each equal to 15$\underline{/20°}$ Ω, are connected between the three supply lines. Express the line voltages and the phase currents in their polar forms.

Solution

The phase voltages are 120$\underline{/50°}$, 120$\underline{/170°}$, and 120$\underline{/-70°}$ volts.

The line voltages are 120$\sqrt{3}\underline{/50 + 30°}$, 120$\sqrt{3}\underline{/170 + 30°}$, 120$\sqrt{3}$ $\underline{/-70 + 30°}$, or 209$\underline{/80°}$, 209$\underline{/-160°}$, and 209$\underline{/-40°}$ volts.

The phase currents are equal to the line currents, which are 209$\underline{/80°}$/ 15$\underline{/20°}$, 209$\underline{/-160°}$/15$\underline{/20°}$, and 209$\underline{/-40°}$/15$\underline{/20°}$, or 13.9$\underline{/60°}$, 13.9$\underline{/180°}$, and 13.9$\underline{/-60°}$ amperes.

THE DELTA (Δ) CONNECTION

Because the sum of the three instantaneous phase voltages is at all times zero, it is possible to connect the phase windings in a delta formation (Chapter 8). This is shown in Fig. 17-7A, and provided the correct connections are made, there will be zero circulating current in the delta loop. Three supply lines then can be connected to the corners of the delta formation, and these will be joined to three loads, also arranged in delta. It is clear that the voltage applied to one of the three loads must be the same as the voltage generated in the corresponding winding across which the load is connected. Therefore in the delta system: line voltage, V_L = phase voltage, V_p.

By contrast, a particular line current, i_X is associated with two phase currents i_{YX} and i_{XZ}. Each of the load currents is then equal to its corresponding phase current. However, each line current is equal to the phasor difference between two of the load currents, so that

$$i_X = i_{XZ} - i_{YX}$$
$$i_Y = i_{YX} - i_{ZY}$$
$$i_Z = i_{ZY} - i_{XZ}$$

The sum of the three line currents is therefore zero; this is true irrespective of whether the loads are balanced or not. However, if the loads are balanced, the phase currents are 120° apart; the relationships between the line and phase currents of the delta system are then comparable to the equations relating to the line and phase voltages in the wye system (Fig. 17-7B). Therefore, in the delta arrangement, line current $I_L = \sqrt{3} \times$ phase current, I_p, and each line current is shifted by 30° from one of the phase currents.

In a balanced wye or delta system, the power in each load is the same, and therefore the total power is

$$P_T = 3 \times E_{rms}(\text{load}) \times I_{rms}(\text{load}) \times \cos \phi$$

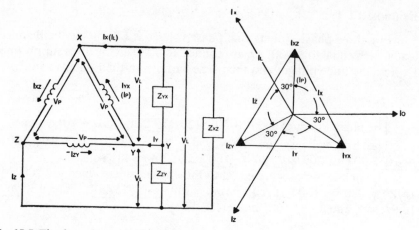

Fig. 17-7 The three-phase delta arrangement.

where $\cos \phi$ is the power factor of each load. For the wye system,

$$E_{rms}(\text{line}) = \sqrt{3} \; E_{rms}(\text{load})$$

and

$$I_{rms}(\text{line}) = I_{rms}(\text{load})$$

Then

$$P_T = 3 \times \frac{E_{rms}(\text{line})}{\sqrt{3}} \times I_{rms}(\text{line}) \times \cos \phi$$

$$= \sqrt{3} \times E_{rms}(\text{line}) \times I_{rms}(\text{line}) \times \cos \phi \; \text{watts}$$

For the delta system

$$E_{rms}(\text{line}) = E_{rms}(\text{load})$$

and

$$I_{rms}(\text{line}) = \sqrt{3} \; I_{rms}(\text{load})$$

Then

$$P_T = 3 \times E_{rms}(\text{line}) \times \frac{I_{rms}(\text{line})}{\sqrt{3}} \times \cos \phi$$

$$= \sqrt{3} \times E_{rms}(\text{line}) \times I_{rms}(\text{line}) \times \cos \phi \; \text{watts}$$

The total power therefore can be measured in terms of the line values, and the expressions are the same for both wye and delta systems.

Example 17-5

In a three-phase delta-connected alternator, the phase voltages are $110\underline{/0°}$ V, $110\underline{/120°}$ V, and $110\underline{/-120°}$ V. Each of the three balanced loads is equal to $5.5\underline{/25°}$ Ω. Calculate the values of the line currents and the total power delivered from the alternator.

Solution

The phase currents are $110\underline{/0°}/5.5\underline{/25°} = 20\underline{/-25°}$ A, $110\underline{/120°}/5.5\underline{/25°} = 20\underline{/95°}$ A, and $110\underline{/-120°}/5.5\underline{/25°} = 20\underline{/-145°}$ A.

The line currents are

$$20\underline{/-25°} - 20\underline{/-145°}$$
$$= 18.1 - j\,8.45 + 16.4 + j\,11.47$$
$$= 34.5 + j\,3.02 = 34.6\underline{/5°}\ \text{A}$$
$$20\underline{/-145°} - 20\underline{/95°}$$
$$= -16.4 - j\,11.47 + 1.47 - j\,19.9$$
$$= -14.66 - j\,31.37 = 34.6\underline{/-115°}\ \text{A}$$

and

$$20\underline{/95°} - 20\underline{/-25°}$$
$$= -1.74 + j\,19.9 - 18.1 + j\,8.45$$
$$= -19.84 + j\,28.35 = 34.6\underline{/125°}\ \text{A}$$

The sum of the line currents is $(34.5 + j\,3.02) + (-14.66 - j\,31.37) + (-19.84 + j\,28.35) = 0$.

Each of the line currents has a magnitude of $20\sqrt{3} = 34.6$ A, and each is shifted by 30° from one of the phase currents.

The total power delivered from the alternator is $\sqrt{3} \times 34.6 \times 110 \times \cos 20° = 6194$ W.

Example 17-6

In a three-phase delta, connected alternator, the phase voltages are $160\underline{/80°}$ V, $160\underline{/-160°}$ V, and $160\underline{/-40°}$ V. The corresponding three unbalanced loads are $20\underline{/40°}$ Ω, $40\underline{/-30°}$ Ω, and $80\underline{/130°}$ Ω. Calculate the values of the line currents.

Solution

The phase currents are

$$\frac{160\underline{/80°}\ \text{V}}{20\underline{/40°}\ \Omega} = 8\underline{/60°}\ \text{A}$$

$$\frac{160\underline{/-160°}\ \text{V}}{40\underline{/-30°}\ \Omega} = 4\underline{/-130°}\ \text{A}$$

$$\frac{160\underline{/-40°}\ \text{V}}{80\underline{/130°}\ \Omega} = 2\underline{/-170°}\ \text{A}$$

The corresponding line currents are

$$8\underline{/60°} - 2\underline{/-170°} = (4 + j\,6.92) - (-1.97 - j\,0.34)$$
$$= 5.97 + j\,7.27$$
$$= 9.41\underline{/50.6°}\ \text{A}$$

$$2\underline{/-170°} - 4\underline{/-130°} = (-1.97 - j\,0.347) - (-2.57 - j\,3.064)$$
$$= 0.60 + j\,2.72$$
$$= 2.79\underline{/77.6°}\ \text{A}$$

and

$$4\underline{/-130°} - 8\underline{/60°} = (-2.57 - j\,3.064) - (4 + j\,6.92)$$
$$= -6.57 - j\,9.99$$
$$= 11.96\underline{/-123.3°}\ \text{A}$$

Although the loads are unbalanced, the total of the line currents is

$$(5.97 + j\,7.27) + (0.60 + j\,2.72) + (-6.57 - j\,9.99) = 0$$

CHAPTER SUMMARY

☐ Revolving Field Alternator

$$\text{generated frequency, } f = \frac{Np}{60}\ \text{Hz}$$

☐ Two-Phase Alternator with Balanced Loads

$$\text{total instantaneous power} = \frac{E_{max}^2}{R}\ \text{watts}$$

$$\text{neutral line current} = \sqrt{2} \times I_{rms}(\text{load})$$

☐ Three-Phase Alternator
The sum of the instantaneous phase voltages is zero.
☐ Wye System with Balanced Loads
Neutral line current is zero.

$$\text{total instantaneous power} = \frac{3\ E_{rms}^2(\text{phase})}{R}\ \text{watts}$$

Line voltages are each $\sqrt{3} \times$ phase voltage and are separated from the phase voltages by 30°.

$$\text{line current} = \text{phase current.}$$
$$\text{total power, } P_T = \sqrt{3} \times E_{rms}(\text{line}) \times I_{rms}(\text{line}) \times \cos\phi\ \text{watts}$$

☐ Delta System
Phasor sum of the line currents is zero.
For balanced loads, line currents are each $\sqrt{3} \times$ phase current and are separated from the phase voltages by 30°.

$$\text{total power, } P_T = \sqrt{3} \times E_{rms}(\text{line}) \times I_{rms}(\text{line}) \times \cos\phi\ \text{watts}$$

18
Motors

IN OUR STUDY of electrical motors, you will learn:

☐ The principles of the dc series motor and its speed versus load and torque versus load characteristics.

☐ The principles of the dc shunt motor and its speed-versus-load and torque-versus-load characteristics.

☐ The use of cumulative and differential compound motors as compromises between the features of dc series and shunt motors.

☐ The principles of the two-phase and three-phase ac induction motors including synchronous speed, rotor speed, and percentage slip.

☐ The operation of split-phase and capacitor single-phase ac induction motors.

☐ The principles of the synchronous ac motor and its ability to provide power-factor correction.

THE SERIES DC MOTOR

In its basic form, a dc motor converts electrical energy into mechanical energy and consists of current-carring conductors that are mounted on a armature and then positioned in a magnetic field. One end of the armature shaft is then connected to the mechanical load. A permanent magnet could provide the necessary flux, but it is more usual to use an electromagnet whose winding is referred to as the *field coil*. Such a coil could be operated separately from another dc source, but it is far more common for the same voltage to drive a current through the low resistance of the armature winding and to excite the field coil.

There are two basic arrangements that have different relationships between the armature speed and a changing mechanical load. In the series-wound motor, the field coil and the armature winding are joined in series with the dc supply voltage, while in the shunt version the field coil and the armature winding are in parallel across the supply.

The circuit of a series-wound motor appears in Fig. 18-1A. When switching on there is initially little resistance to limit the current so that there is a strong magnetic field and a high starting torque. As the armature speeds up, its counter EMF increases until an equilibrium condition is reached for a particular load. The rheostst, R, normally is included with high power motors to reduce the initial current and to control the speed to a limited extent.

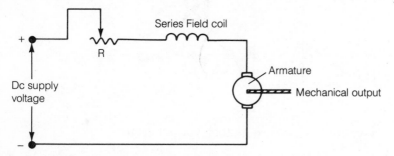

Fig. 18-1 A Mechanical output.

If the load on the motor increases, the armature slows down and the counter EMF is reduced. The current through the armature is greater and therefore the field strength increases; the speed is then reduced to a low level. However, the armature current is not excessive because the torque is directly proportional to the product of the flux density and the armature current. Consequently, the torque varies as the square of the current. The speed regulation of a series-wound motor is poor and its percentage speed regulation, as defined below, is high (Fig. 18-1B):

$$\text{percentage speed regulation} = \frac{N_{NL} - N_{FL}}{N_{FL}} \times 100\%$$

where

N_{NL} = speed under no-load or minimum load conditions in revolutions per minute (rpm)

N_{FL} = speed under full-load conditions (rpm)

Disaster occurs if the load is removed abruptly from the series-wound motor; the armature speeds up and a higher counter EMF is induced into the armature winding. This reduces the armature current that also flows through the field coil. The weakened field causes the armature to turn still faster. The effect is therefore cumulative, and the motor ultimately will be destroyed. For this reason, series motors are connected directly to their loads and never by a belt.

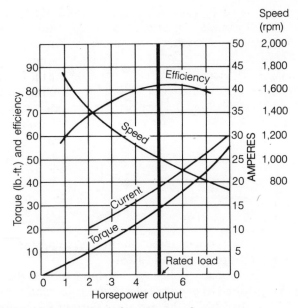

Fig. 18-1 B Circuit and characteristics of a series-wound motor.

Summarizing, the series-wound motor has as excellent starting torque but poor speed regulation.

Example 18-1

The minimum load speed of a dc motor is 2000 rpm while on full load the speed drops to 1800 rpm. Calculate the value of the speed regulation.

Solution

$$\text{percentage of speed regulation} = \frac{(2000 - 1800) \times 100\%}{1800}$$

$$= 11\%$$

THE SHUNT DC MOTOR

The basic circuit of a shunt wound motor appears in Fig. 18-2A. Initially, the rheostat in series with the field coil is adjusted for zero resistance so that the magnetic flux is at its highest level. The rheostat in series with the armature is a starter box whose resistance is set initially at its maximum value so that the armature current is limited to a safe level. As the armature speeds up to create the counter EMF, the resistance of the starter box is cut out in a series of steps. When this resistance finally reaches zero, the armature runs at its full speed corresponding to the load; the sum of the counter EMF and IR drop across the armature is then equal to the source voltage. Subsequently, the speed can be controlled by setting the field rheostat. For exam-

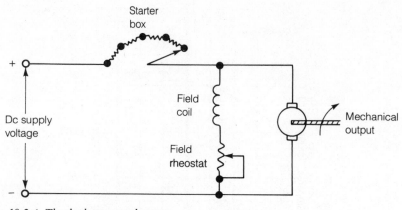

Fig. 18-2 A The dc shunt-wound motor.

ple, if the rheostat's resistance increases, the field is reduced and the armature speed increases to restore the counter EMF.

When there is no load applied to the shunt motor, the only torque necessary is that required to overcome bearing friction and wind resistance. The counter EMF then limits the armature current to the low level that provides the necessary torque to run the motor without a load. However, if the field coil then opens or burns out, the counter EMF nearly drops to zero; the armature then races at an ever increasing speed until the motor is destroyed.

When the external load is applied, the shunt motor slows down only slightly. The small decrease in speed causes a corresponding fall in the counter EMF. However, because the armature resistance is low, the increases in the armature current and the torque are relatively large. The torque rises until it provides the value required by the load. The speed then stablizes at a new value that is determined by the load. Because wide variations in the load cause only small changes in the shunt motor's speed, its speed regulation is good; Fig. 18-2B illustrates this fact.

The mechanical output of a dc motor is rated in horsepower where 1 horsepower (hp) = 746 watts. The equation is

$$\text{output power} = \frac{E \times I \times F}{746 \times 100} \text{ hp}$$

where

$$E = \text{dc source voltage (V)}$$
$$I = \text{dc source current (A)}$$
$$F = \text{percentage efficiency (\%)}$$

Compound motors (Figs. 18-3A and B) are an attempt to improve on speed/load and the torque/load characteristics. Such motors have both a series- and a shunt-field coil, and if the two fields are aiding, the motor is of the cumulative compound type. By contrast, the fields are opposing in the differential compound motor.

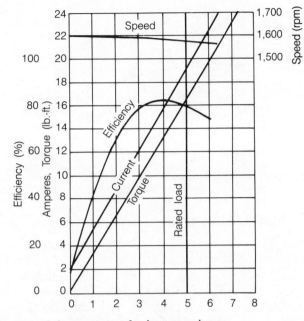

Fig. 18-2 B Circuit and characteristics of a shunt-wound motor.

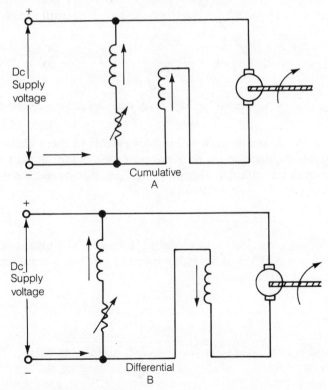

Fig. 18-3 Examples of compound motors.

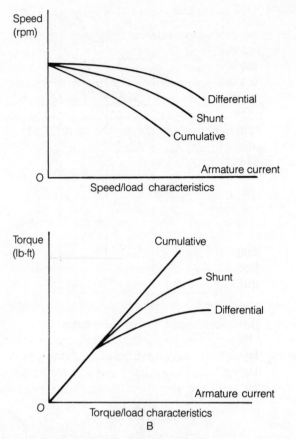

Fig. 18-4 Comparison between the characteristics of shunt and compound motors.

Figure 18-4A and B show the improvements in these characteristics. The cumulative compound variety provides greater torque than either the shunt motor or the differential compound type, but its speed regulation is inferior.

Example 18-2

A dc motor is supplied with a source voltage of 220 V and a corresponding source current of 5 A. If the motor efficiency is 80%, what is the motor's mechanical output in hp?

Solution

$$\text{output power} = \frac{220 \text{ V} \times 5 \text{ A} \times 80}{746 \times 100}$$

$$= 1.2 \text{ hp}$$

THE INDUCTION MOTOR

For three-phase operation, the induction motor has a starter that contains three windings with their axes separated by 120° from each other (Fig. 18-5A). When these wye-connected windings are excited by a three-phase supply, the result is a two-pole magnetic field whose rotation is synchronized with the frequency of the source (Fig. 18-5B). For example, if the line frequency is 60 Hz, the synchronous speed of the rotating magnetic field is $60 \times 60 = 3600$ rpm and is independent of the motor's load. However, if the number of the stator windings is doubled to create a four-pole motor, the synchronous speed is halved. In equation form,

$$N_S = \frac{60\,f}{p} \text{ rpm}$$

where

N_S = magnetic field's synchronous speed of rotation
f = frequency of the three phase supply (Hz)
p = number of pole pairs

A "squirrel" rotor (Fig. 18-5C) revolves inside the stator and has a laminated iron cylindrical core with parallel slots, in which copper bars are imbedded. These bars are connected together at each end by copper rings and are not insulated from the core because the induced currents follow paths of least resistance. It is theoretically possible to use a solid iron rotor, but that would create large eddy current losses; such a rotor would severely overheat.

The revolving field produced by the stator windings cuts the rotor bars in which voltages are therefore induced. Currents then flow in these conductors because the end-rings provide closed electrical paths. The direction of the resulting torque is such as to drag the rotor in the same direction as that of the rotating field (Fig. 18-5D). The torque's magnitude is proportional to the product of the rotor current and the field strength.

Under no-load conditions, the rotor accelerates until its speed, N_r, approaches the synchronous speed, N_s. However, in order to provide a small torque to overcome friction and windage, there must always be some relative motion between the rotor and the revolving field so that N_r is less than N_s. This difference between the two speeds is expressed in terms of the percentage slip, S, defined by

$$\text{percentage slip, } S = \frac{N_s - N_r}{N_s} \times 100\%$$

Normally the value of S is about 1% to 2% for no-load conditions. When the normal load is applied, the slip must increase (typically 5% to 10%) in order to provide a greater torque.

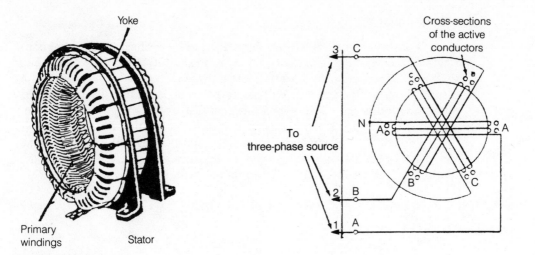

Yoke

Primary
windings   Stator

Fig. 18-5 A Stator.

Cross-sections
of the active
conductors

To
three-phase source

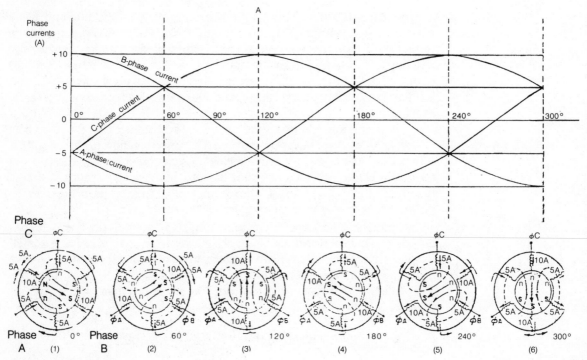

Fig. 18-5 B Phase currents.

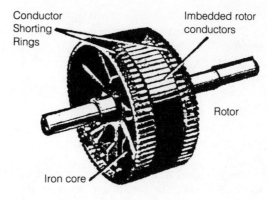

Conductor Shorting Rings

Imbedded rotor conductors

Rotor

Iron core

C

Fig. 18-5 C Rotor.

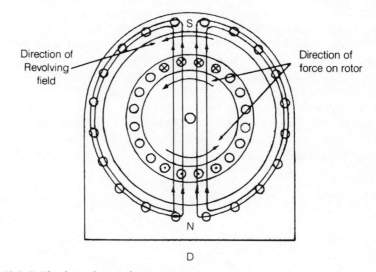

Direction of Revolving field

Direction of force on rotor

D

Fig. 18-5 D The three-phase induction motor.

Because of reactive effects, the induction motor has a lagging power factor whose value depends on the load. The motor's mechanical output is then given by

$$\text{output power} = \frac{\text{apparent input power} \times \text{power factor} \times \text{efficiency\%}}{746 \times 100} \text{ hp}$$

The typical efficiency for an induction motor is about 90%.

For two phase operation, the induction motor has a stator with two windings whose axes are perpendicular. These windings are fed with two currents that are 90° out of phase; the result again is a magnetic field that revolves in synchronization with the supply cycle. It is also possible to operate an induction motor with a cage rotor from a single-phase supply. When the rotor is running at the normal speed corresponding to its load, the action is similar to that of three-phase operation. However, when the rotor is initially at rest, there is no revolving field, and therefore the starting torque is zero. However, if the rotor can be brought up to speed by separate means, the induced currents in the rotor bars combine with the stator currents to produce a revolving field that causes the rotor to continue to run in the direction in which it started.

One method of providing a single-phase motor with a starting torque is to create the effect of two-phase operation. In a *split-phase motor* (Fig. 18-6A), the stator contains a main winding for running conditions and an auxiliary winding to provide the starting torque; the axes of the two windings are 90° apart. The main winding has a higher reactance and a lower resistance than the auxilary winding so that the two currents are shifted in phase by approximately 15°. This is sufficient to produce a magnetic field that revolves at synchronous speed and creates the starting torque. When the rotor has accelerated to about 75% of the synchronous speed, a centrifugally operated switch disconnects the starting winding from the line voltage, and the motor continues to run on the main winding along.

In another solution, a capacitor of several microfarads is connected in series with the auxiliary winding, which has more turns of thicker wire than the corresponding winding of the split-phase motor. Because the phase difference between the two currents is closer to 90° (Fig. 18-6B), the starting torque of the "capacitor" motor is greater and can be as much as 350% of the full-load torque.

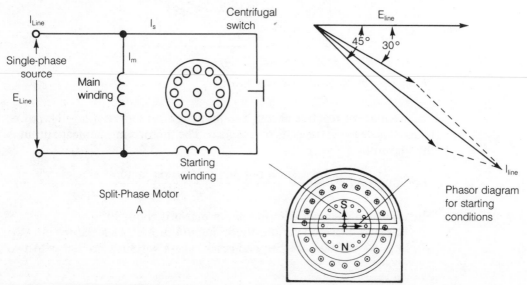

Fig. 18-6 A Split-phase motor.

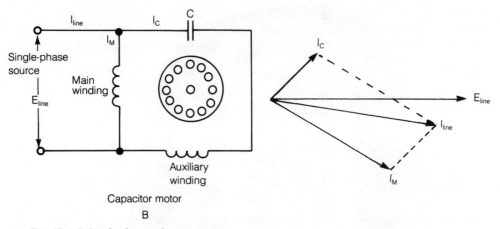

Fig. 18-6 B Single-phase induction motors.

Example 18-3

On no load, the speed of a three-phase, two-pole, 60-Hz motor is 3500 rpm, while on full load the speed drops to 3300 rpm. What is the value of the motor slip on (a) no load and (b) full load?

Solution

$$\text{(a) synchronous speed} = \frac{60 \times 60}{1} = 3600 \text{ rpm}$$

$$\text{no-load rotor slip} = \frac{3600 - 3500}{3600} \times 100 = 2.8\%$$

$$\text{(b) full-load rotor slip} = \frac{3600 - 3300}{3600} \times 100 = 8.3\%$$

THE SYNCHRONOUS MOTOR

The stator windings of the synchronous motor are virtually the same as those of the induction motor. When three-phase currents are supplied to the stator windings of the synchronous motor (Fig. 18-7A), the rotating field created is the same as in the induction motor. However, the rotor is now in the form of an electromagnet that is supplied from a dc source through slip rings (Fig. 9-7B); the result is a fixed polarity at each rotor pole. A rheostat in series with the dc source controls the strength of the rotor's field.

If we assume that the rotor has no inertia and there is no friction or windage, under no-load conditions the rotor revolves instantaneously in step with the rotating field as soon as the stator and rotor windings are excited. However, in practice, the starting torque is zero (Fig. 18-7C) and the rotor must be brought up to the synchronous speed by separate means. One common method is to include a cage winding on the rotor so that the

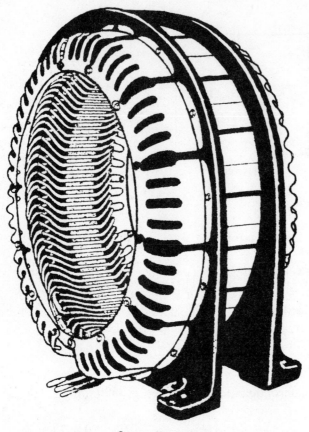

Stator   A

Fig. 18-7 A Stator.

machine is self-starting as an induction motor. Initially, the dc source is removed and the field rheostat is set to its minimum current position. The rotor is then accelerated to a speed that is slightly below the synchronous value. When the dc source is reconnected and the excitation current is increased, the rotor locks into step with the stator's revolving field.

For a given load, the power factor of a synchronous motor is dependent on the rotor's exciting current. It is possible to vary the power factor from lagging through unity to leading, and consequently a synchronous motor under no load is frequently used for power-factor correction. The motor is placed in parallel with a lagging inductive load, and the rotor's excitation is then adjusted until the power factor of the combination is unity. The line current is then reduced to its minimum level at the expense of the small amount of true power taken by the synchronous motor under no-load conditions. When the motor is used in this way to draw a leading current, it is sometimes referred to as a *synchronous capacitor*.

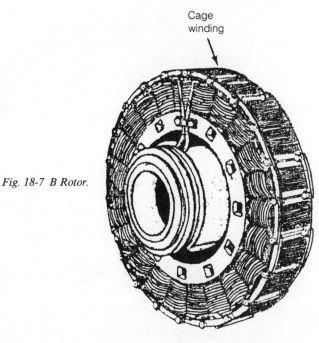

Fig. 18-7 B Rotor.

Rotor B

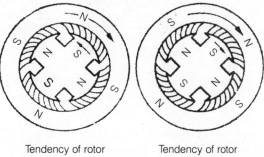

Fig. 18-7 C Construction and principle of the synchronous motor.

Tendency of rotor to turn counter-clockwise

Tendency of rotor to turn clockwise

C

Example 18-4

A synchronous motor has a source voltage of 220 V and a total line current of 15 A. If the motor's power factor is 0.85 leading and its efficiency is 80%, calculate the value of the motor's output horsepower.

Solution

$$\text{power output} = \frac{220 \text{ V} \times 15 \text{ A} \times 0.85 \times 80}{746 \times 100}$$

$$= 3 \text{ hp}$$

CHAPTER SUMMARY

☐ *Percentage (Speed) Regulation of dc Motors*

$$= \frac{N_{NL} - N_{FL}}{N_{FL}} \times 100\%$$

$$\text{mechanical power output of dc Motors} = \frac{E \times I \times F}{746 \times 100} \text{ hp}$$

☐ *Synchronous Speed of the Induction Motor*

$$N_S = \frac{60 \text{ f}}{p} \text{ rpm}$$

$$\text{percentage slip, } S = \frac{N_s - N_r}{N_r} \times 100\%$$

Mechanical power output

$$= \frac{\text{apparent input power} \times \text{power factor} \times \text{efficiency } \%}{746 \times 100} \text{ hp}$$

Power factor of the induction motor is lagging.

☐ *Speed of the Synchronous Motor*

$$\text{speed} = \frac{60 \text{ f}}{p} \text{ rpm}$$

Power factor of the synchronous motor is either leading or unity or lagging.

19

Nonsinusoidal Waveforms

As WE INVESTIGATE the properties of nonsinusoidal waveforms such as the square wave, the sawtooth and the pulse, you will learn

☐ How nonsinusoidal waveforms can be synthesized by combining a number of sinewaves and cosine waves.

☐ About the Fourier series and its ability to express finite, continuous, and single-valued functions in terms of sine and cosine series.

☐ About the harmonic content of the Fourier expressions, which represent a number of nonsinusoidal waveforms.

☐ How to calculate the effective value of a nonsinusoidal waveform.

☐ How to determine the expression for the nonsinusoidal current in a series ac circuit.

☐ How to determine the expression for the total nonsinusoidal current in a series-parallel ac circuit.

In previous chapters we have been concerned only with sinusoidal voltages and currents. However, what about other alternating waveforms such as the sawtooth and the triangle wave (see Fig. 19-1A – G)? Do we require a new method of analysis for each type of alternating source? Fortunately we do not. In the eighteenth century, a French mathematician by the name of Jean Baptiste Fourier (1768 – 1850) was trying to solve a problem involving the flow of heat. He developed a means of analysis by which any finite, continuous, and periodic waveform could be analyzed into a series of sinewaves. The general series contains a fundamental sinewave whose frequency is the same as that of the periodic waveform; together with the fundamental wave are harmonic components so that the derivation of the Fourier series is sometimes known as *harmonic analysis*.

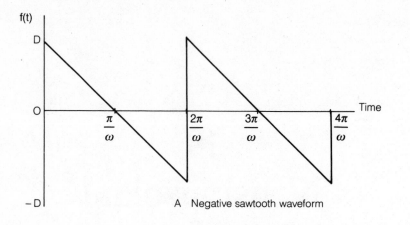

A   Negative sawtooth waveform

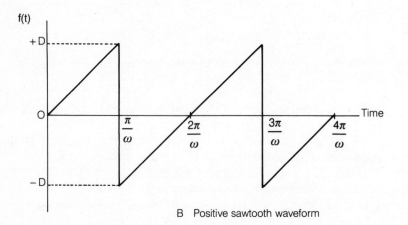

B   Positive sawtooth waveform

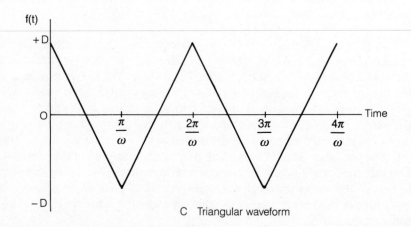

C   Triangular waveform

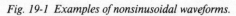

Fig. 19-1 Examples of nonsinusoidal waveforms.

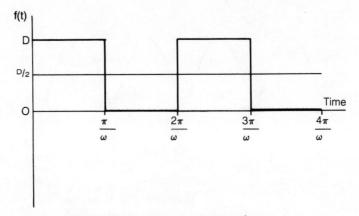
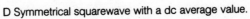

D Symmetrical squarewave with a dc average value.

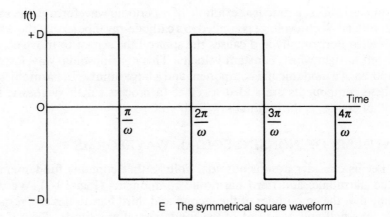

E   The symmetrical square waveform

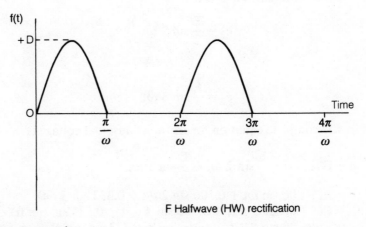

F Halfwave (HW) rectification

Fig. 19-1 Continued.

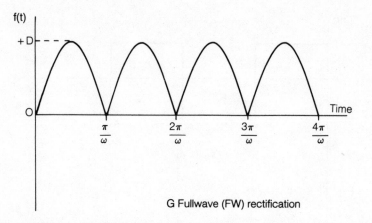

G Fullwave (FW) rectification

Fig. 19-1 Continued.

In electronics, a practical example of a periodic waveform is the saw-tooth voltage. Such a voltage is applied to a cathode-ray tube so that the beam is deflected horizontally and causes the spot on the screen to move across from left to right with a constant velocity. This nonsinusoidal waveform is composed of a fundamental component and a large number of harmonics. If all these components are added together (a process called *synthesis*) the result, of course, is the original sawtooth voltage.

SYNTHESIS OF NONSINUSOIDAL WAVEFORMS

Let us consider a nonsinusoidal voltage that contains fundamental, second harmonic, and third harmonic components (Fig. 19-2). We will assume that the amplitudes of the second and third harmonics are, respectively, one-half and one-third of the fundamental amplitude. Then if the fundamental sinewave is represented by $e_1 = E \sin \omega t$, the expressions for the second and third harmonics are

$$e_2 = \frac{E}{2} \sin 2 \omega t$$

and

$$e_3 = \frac{E}{3} \sin 3 \omega t$$

Therefore the complete equation for the nonsinusoidal voltage is

$$e = E \sin \omega t + \frac{E}{2} \sin 2 \omega t + \frac{E}{3} \sin 3 \omega t$$

$$= E(+1.00 \sin \omega t + 0.500 \sin 2 \omega t + 0.333 \sin 3 \omega t)$$
$$= E(+1.00 \sin 2 \pi ft + 0.500 \sin 4 \pi ft + 0.333 \sin 6 \pi ft)$$
$$= E(+1.00 \sin \theta + 0.500 \sin 2 \theta + 0.333 \sin 3 \theta)$$

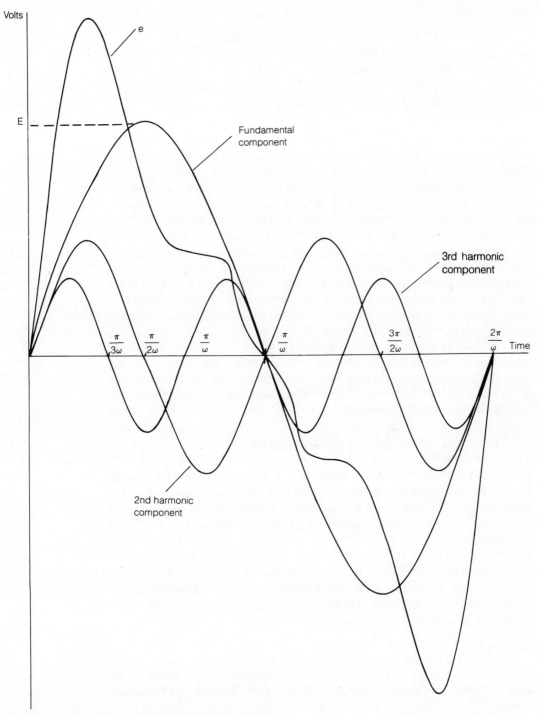

Fig. 19-2 Synthesis of a nonsinusoidal voltage waveform.

where

> e = instantaneous value of the nonsinusoidal wave (V)
> E = amplitude of the fundamental sine wave (V)
> ω = angular frequency (rad/s)
> f = frequency (Hz)
> θ = angle (rad)
> t = time (s)

Note that the three components (fundamental and harmonics) start off together at the beginnings of their positive half cycles. This accounts for the positive signs in front of all the terms in the equation for the voltage.

To derive the nonsinusoidal wave we plot the fundamental and the two harmonics on a common graph by using a horizontal radian (or degree) scale. At a number of conveniently chosen points we then take the algebraic sum of the vertical distances associated with the instantaneous values of the three components.

When we study the resultant or e waveform, we are struck with its similarity to a sawtooth voltage with a negative slope (Fig. 19-1A). In fact if we add more and more harmonics (of the correct size), we obtain closer and closer approximations to the negative sawtooth waveform. This is well illustrated by the waveforms of Fig. 19-3.

The perfect sawtooth can be obtained only by including an infinite number of harmonics. This would lead you to question the value of Fourier analysis. However we should note that the higher the order of the harmonic, the less is its amplitude. In our example of the negative sawtooth, the amplitude of the hundredth harmonic

$$(e_{100} = \frac{E}{100} \sin 100 \ \omega t)$$

is only 1% of the fundamental amplitude. It follows that only a limited number of harmonics are required to obtain a good approximation to the negative sawtooth voltage. When sufficient harmonics are included, the peak value of the sawtooth waveform is approximately $1.57 \times E$ volts (quoted result).

Summarizing, we can obtain nonsinusoidal ac waveforms by combining fundamental and harmonic components. Theoretically, the Fourier series contains an infinite number of harmonics, but in practice only a limited number are necessary in order to obtain a good approximation to the required waveform.

Example 19-1

The e waveform of Fig. 19-2 has a fundamental component whose amplitude is 6 V. Calculate the amplitudes of the second and third harmonics.

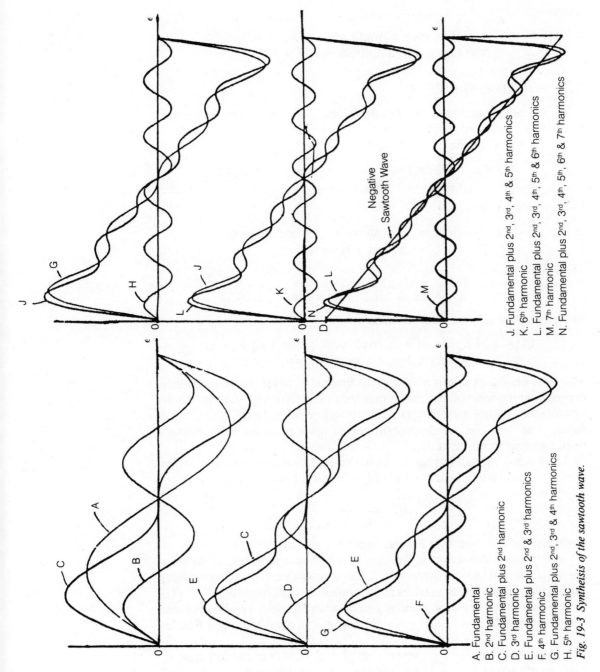

A. Fundamental
B. 2nd harmonic
C. Fundamental plus 2nd harmonic
D. 3rd harmonic
E. Fundamental plus 2nd & 3rd harmonics
F. 4th harmonic
G. Fundamental plus 2nd, 3rd & 4th harmonics
H. 5th harmonic

J. Fundamental plus 2nd, 3rd, 4th & 5th harmonics
K. 6th harmonic
L. Fundamental plus 2nd, 3rd, 4th, 5th & 6th harmonics
M. 7th harmonic
N. Fundamental plus 2nd, 3rd, 4th, 5th, 6th & 7th harmonics

Negative Sawtooth Wave

Fig. 19-3 Syntheisis of the sawtooth wave.

Solution

$$\text{second harmonic amplitude} = 6/2 = 3 \text{ V.}$$
$$\text{third harmonic amplitude} = 6/3 = 2 \text{ V.}$$

THE FOURIER SERIES

The Fourier theorem can be stated formally as follows: any finite, continuous, single-valued function, f(t), that has a period of 2π radians (360°), can be expressed by the following series:

$$f(t) = a_0 + A_1 \sin(\omega t + \phi_1) + A_2 \sin(2\omega t + \phi_2)$$
$$+ A_3 \sin(3\omega t + \phi_3) + \dots$$

It can be shown that

$$A \sin(\omega t + \phi) = A \sin \omega t \cos \phi + A \cos \omega t \sin \phi$$
$$= a \cos \omega t + b \sin \omega t$$

where

$$a = A \sin \phi \quad \text{and} \quad b = A \cos \phi$$

Therefore the series can be restated as

$$f(t) = a_0 + a_1 \cos \omega t + a_2 \cos 2\omega t + a_3 \cos 3\omega t + \dots$$
$$+ b_1 \sin \omega t + b_2 \sin 2\omega t + b_3 \sin 3\omega t + \dots$$

Now what does all this mean? In the first place, f(t) is the mathematical expression for a function of time, or in other words, f(t) is any quantity whose instantaneous value is dependent on the time. For example, $f(t) = \sin \omega t$ is a function of time because the instantaneous value of a sine wave changes from moment to moment.

The word "finite" indicates that f(t) contains no infinities because the sum of a Fourier series can never tend to infinity. For example,

$$f(t) = \frac{1}{t - 1}$$

could not be expressed by a Fourier series because as $t \to 1$, $f(t) \to \infty$.

Single-valued means that f(t) cannot have more than one value for a particular time. This is perhaps obvious because the sum of a Fourier series cannot have two different answers at the same time. The function $f(t) = \sqrt{t}$ is double valued because each positive real number has both a positive square root and a negative root; such a function cannot be analyzed into a Fourier series.

Continuous means that there are no discontinuities in the function. If, for example, $f(t) = +\sqrt{(t-1)(t-2)}$, f(t) has no real value between t = 1 and t = 2 (because we cannot take the square root of a negative number), and clearly the sum of the Fourier series never can be an imaginary quantity.

Turning to the series itself, the term "a_0" represents the mean level of f(t). If the positive and negative excursions of a nonsinusoidal waveform are not equal, the waveform can be regarded as composed of a dc value (a_0) together with the alternating components ($A_1 \sin(\omega t + \phi_1)$, $A_2 \sin(2\omega t + \phi_2)$ and so on). Examples are the half- and full-wave rectified ac voltages of Figs. 19-1F and G.

Note that the fundamental and harmonic components are each associated with the phase angles, ϕ_1 and ϕ_2, etc. These phase angles were zero in our example of the negative sawtooth, but in general the existence of these phase angles means that the fundamental and the harmonics do not start off together at the beginnings of their positive half cycles.

In order to use the series to analyze a complex waveform, you must determine the coefficients $a_1, a_2, \ldots$, and $b_1, b_2, \ldots$. You do this by multiplying both sides of the Fourier equation by a suitable factor and then integrating between the limits 0 and 2π. If the multiplying factor is chosen correctly, all the terms vanish except those that give the required coefficient. This method leads to the following results:

$$a_0 = \frac{1}{2\pi} \times \int_0^{2\pi} f(t)\, d(\omega t)$$

Note that a_0 is the mean value of $f(t)$ between the limits 0 and 2π.

$$a_n = \frac{1}{\pi} \times \int_0^{2\pi} f(t) \cos n\, \omega t\, d(\omega t)$$

where n is any positive integer

The value of a_n is twice the mean value of $f(t) \cos n\, \omega t$ between the limits of 0 and 2π.

$$b_n = \frac{1}{\pi} \times \int_0^{2\pi} f(t) \sin n\, \omega t\, d(\omega t)$$

where n is any positive integer

Therefore b_n is twice the mean value of $f(t) \sin n\, \omega t$ between the limits of 0 and 2π.

Let us find the values of the coefficients in the Fourier analysis of the square waveform shown in Fig. 19-1D.

The squarewave of Fig. 19-1D is a single-valued finite, continuous periodic function of ωt and has a period of 2π; it therefore can be analyzed by Fourier's Theorem.

From $t = 0$ to $t = \pi/\omega$, the equation of the function is $f(t) = D$.

From $t = \pi/\omega$ to $t = 2\pi/\omega$, the equation of the function is $f(t) = 0$. Then

$$a_0 = \frac{1}{2\pi} \int_0^{2\pi} f(t)\, d(\omega t) = \frac{1}{2\pi} \int_0^{\pi} f(t)\, d(\omega t) + \frac{1}{2\pi} \int_\pi^{2\pi} f(t)\, d(\omega t)$$

$$= \frac{1}{2\pi} \times D \times \pi + 0 = \frac{D}{2}$$

This verifies that the mean value of f(t) is D/2. Also

$$a_n = \frac{1}{\pi} \times \int_0^{2\pi} f(t) \cos n \, \omega t \, d(\omega t)$$

$$= \frac{1}{\pi} \times \left[\int_0^{\pi} D \cos n \, \omega t \, d(\omega t) + \int_{\pi}^{2\pi} 0 \cos n \, \omega t \, d(\omega t) \right]$$

$$= \frac{1}{\pi} \times \left[\frac{D \sin n \, \omega t}{n} \right]_0^{\pi} = 0$$

All cosine terms are therefore zero. This is because the symmetry of the waveform.

Finally,

$$b_n = \frac{1}{\pi} \times \int_0^{2\pi} f(t) \sin n \, \omega t \, d(\omega t)$$

$$= \frac{1}{\pi} \times \left[\int_0^{\pi} D \sin n \, \omega t \, d(\omega t) + \int_{\pi}^{2\pi} 0 \sin n \, \omega t \, d(\omega t) \right]$$

$$= \frac{1}{\pi} \times \left[\frac{-D \cos n \, \omega t}{n} \right]_0^{\pi}$$

$$= \frac{D}{n \, \pi} (1 - \cos n \, \pi)$$

When n is odd, $(1 - \cos n \, \pi) = 2$
When n is even, $(1 - \cos n \, \pi) = 0$

This yields

$$b_1 = \frac{2 \, D}{\pi}, \ b_2 = 0, \ b_3 = \frac{2 \, D}{3 \, \pi}, \ b_4 = 0 \quad \text{and so on}$$

The required equation for the square wave is

$$f(t) = \frac{D}{2} + \frac{2 \, D}{\pi} \left[\sin \omega t + \frac{1}{3} \sin 3 \, \omega t + \frac{1}{5} \sin 5 \, \omega t + \frac{1}{7} \sin 7 \, \omega t + \ldots \right]$$

A waveform frequently encountered in communications appears in Fig. 19-1E. This is similar to the one shown in Fig. 19-1D, which was represented by the above equation. The new waveform, however, is now symmetrical about the time axis. Therefore the first term, D/2, does not appear in its equation, which is

$$f(t) = \frac{2 \, D}{\pi} \times \left[\sin \omega t + \frac{1}{3} \sin 3 \, \omega t + \frac{1}{5} \sin 5 \, \omega t + \ldots \right]$$

$$= E \, (+1.00 \sin \theta + 0.333 \sin 3 \, \theta + 0.200 \sin 5 \, \theta \ldots)$$

The synthesis of this square wave is illustrated in Fig. 19-4; its peak value, D/2, is $\pi \times E/4 = 0.785E$ approximately.

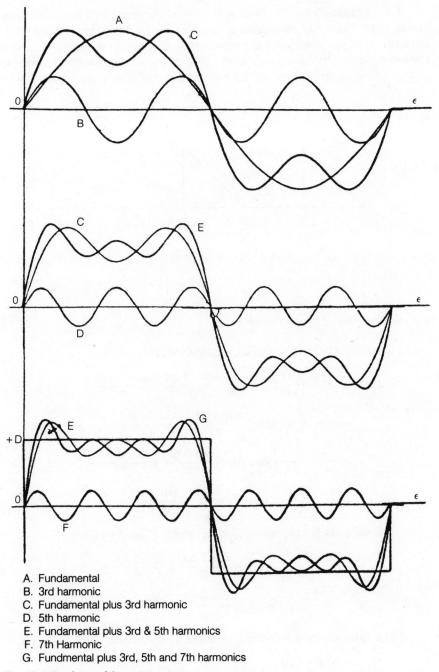

A. Fundamental
B. 3rd harmonic
C. Fundamental plus 3rd harmonic
D. 5th harmonic
E. Fundamental plus 3rd & 5th harmonics
F. 7th Harmonic
G. Fundmental plus 3rd, 5th and 7th harmonics

Fig. 19-4 Synthesis of the square wave.

The complete Fourier series is shown as composed of both sine and cosine terms (you will recall that a cosine wave leads a sinewave by 90°). Because of the symmetries of the waveforms, there were no cosine terms in the two series for the sawtooth and the symmetrical square wave. However, in the following series there are a number of cosine terms (see Figure 19-5):

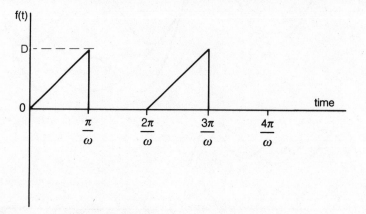

Fig. 19-5 The modified sawtooth waveform.

The formula for a modified sawtooth waveform is:

$$f(t) = \frac{D}{4} - \frac{2D}{\pi^2}\left(\cos \omega t + \frac{\cos 3\ \omega t}{3^2} + \frac{\cos 5\ \omega t}{5^2} + \ldots\right)$$

$$+ \frac{D}{\pi}\left(\sin \omega t - \frac{\sin 2\ \omega t}{2} + \frac{\sin 3\ \omega t}{3} + \ldots\right)$$

$$= \frac{D}{4} - \frac{2D}{\pi^2}\left(\cos \omega t + \frac{\cos 3\ \omega t}{9} + \frac{\cos 5\ \omega t}{25} + \ldots\right)$$

$$+ \frac{D}{\pi}\left(\sin \omega t - \frac{\sin 2\ \omega t}{2} + \frac{\sin 3\ \omega t}{3} + \ldots\right)$$

The formula for the triangular waveform (Fig. 19-1C) is:

$$f(t) = \frac{8D}{\pi^2}\left(\cos \omega t + \frac{\cos 3\ \omega t}{3^2} + \frac{\cos 5\ \omega t}{5^2} + \ldots\right)$$

$$= \frac{8D}{\pi^2}\left(\cos \omega t + \frac{\cos 3\ \omega t}{9} + \frac{\cos 5\ \omega t}{25} + \ldots\right)$$

The formula for a half-wave rectified ac waveform (Fig. 19-1F) is:

$$f(t) = \frac{2D}{\pi}\left(\frac{1}{2} + \frac{\pi}{4}\sin \omega t - \frac{1}{1 \times 3}\cos 2\ \omega t - \frac{1}{3 \times 5}\cos 4\ \omega t\right.$$

$$\left. - \frac{1}{5 \times 7}\cos 6\ \omega t - \ldots\right)$$

$$= \frac{2\,D}{\pi} \left(\frac{1}{2} + \frac{\pi}{4} \sin \omega t - \frac{1}{3} \cos 2\,\omega t - \frac{1}{15} \cos 4\,\omega t \right.$$

$$\left. - \frac{1}{35} \cos 6\,\omega t - \ldots \right)$$

The formula for a full-wave rectified ac waveform (Fig. 19-1G) is

$$f(t) = \frac{4\,D}{\pi} \left(\frac{1}{2} - \frac{1}{1 \times 3} \cos 2\,\omega t - \frac{1}{3 \times 5} \cos 4\,\omega t - \frac{1}{5 \times 7} \cos 6\,\omega t \right.$$

$$= \frac{4\,D}{\pi} \left(\frac{1}{2} - \frac{1}{3} \cos 2\,\omega t - \frac{1}{15} \cos 4\,\omega t - \frac{1}{35} \cos 6\,\omega t \ldots \right)$$

In each of the expressions, D is the peak value of the waveform.

The square wave, sawtooth, and reflected ac waveforms are encountered frequently in electronics. The pulse waveform also is quite common, but its analysis is beyond the scope of this book.

To summarize, we have examined a number of periodic, finite, single-valued and continuous waveforms whose Fourier expansions in general contain a constant value together with sine and cosine terms that represent the fundamental and harmonic components.

Example 19-2

The half-wave rectified ac waveform of Fig. 19-1F has a peak value of 100 V. Calculate the amplitudes of the fundamental component and of all the harmonics up to and including the sixth.

Solution

The mean value or dc level is

$$a_0 = 100 \text{ V} \times \frac{1}{\pi} = 31.8 \text{ V}$$

The amplitude of the fundamental sinewave component is

$$b_1 = 100 \text{ V} \times \frac{1}{2} = 50 \text{ V}$$

The amplitude of the second harmonic component is

$$a_2 = (-) \, 100 \text{ V} \times 2/(3\,\pi) = (-) \, 21.2 \text{ V}$$

The amplitude of the fourth harmonic component is

$$a_4 = (-) \, 100 \text{ V} \times 2/(5\,\pi) = (-) \, 4.24 \text{ V}$$

The amplitude of the sixth harmonic component is

$$a_6 = (-) \, 100 \text{ V} \times 2/(35\,\pi) = (-) \, 1.83 \text{ V}$$

EFFECTIVE (RMS) VALUE OF A NONSINUSOIDAL WAVE

We already have seen that a nonsinusoidal wave in general can be expressed in terms of a dc level together with fundamental and harmonic components. Let us assume that this wave is in a form of a source voltage that is connected across a resistor, R. Each component provides its own power dissipation and because the individual powers dissipated are additive, we can derive an equation for the effective voltage of the source. This value would be the reading of an ac voltmeter connected across the source provided the instrument can respond to the frequencies of the higher harmonics.

$$\text{total power dissipated, } P_T = \frac{V_{dc}^2}{R} + \frac{V_1^2}{R} + \frac{V_2^2}{R} + \ldots \text{ watts}$$

where

$$V_{DC} = \text{dc level of the voltage source (V)}$$
$$V_1 = \text{effective value of the fundamental component (V)}$$
$$V_2 = \text{effective value of the second harmonic component (V)}$$

But

$$P_T = V_T^2/R \text{ watts}$$

where V_T = effective value of the voltage source (V)

Therefore

$$V_T^2 = V_{dc}^2 + V_1^2 + V_2^2 + \cdots$$

or

$$V_T = \sqrt{V_{dc}^2 + V_1^2 + V_2^2 + \cdots} \text{ volts}$$

A similar analysis reveals that

$$I_T = \sqrt{I_{dc}^2 + I_1^2 + I_2^2 \cdots} \text{ amperes}$$

where I_T is the effective value of the nonsinusoidal current.

Example 19-3

Calculate the total effective value of the component voltages obtained in Example 19-2.

Solution

$$\text{total effective voltage} = \sqrt{31.8^2 + \frac{50^2}{2} + \frac{21.2^2}{2} + \frac{4.24^2}{2} + \frac{1.83^2}{2}}$$
$$\approx 50 \text{ V}$$

This shows that the rms value of the half sine wave is 0.5 of the peak value. By comparison, the RMS value of the full sine wave is 0.707 of the peak value.

NONSINUSOIDAL CURRENT IN A SERIES AC CIRCUIT

If a nonsinusoidal voltage source is applied across a series circuit, the resulting current contains fundamental and harmonic components provided all these components are contained in the voltage waveform.

When the series ac circuit contains both inductors and capacitors, their reactances must be calculated separately for the fundamental and harmonic frequencies in order to obtain the values of the corresponding currents. In other words, for each frequency the current must be recalculated. When the values of all the current components are known, we can obtain the effective value of the nonsinusoidal current drawn from the source. I will illustrate these principles by the following example.

A nonsinusoidal source has a voltage that is represented by the equation $e = 100 \sin \omega t - 80 \sin (2\,\omega t + 40°) + 40 \sin (3\,\omega t - 20°)$ V, where the frequency $f = \omega/2\,\pi = 400$ kHz. This voltage then is applied across the series circuit of Fig. 19-6. We will now obtain the equation of the nonsinusoidal current and then calculate its effective value.

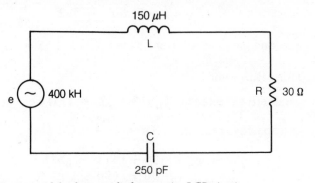

Fig. 19-6 Nonsinusoidal voltage applied to a series LCR circuit.

Step 1. The Fundamental Component

$$\text{inductive reactance, } X_L = 2 \times \pi \times f \times L$$
$$= 2 \times \pi \times 400 \times 10^3 \times 150 \times 10^{-6}$$
$$= 377 \; \Omega$$

$$\text{capacitive reactance, } X_C = \frac{1}{2 \times \pi \times f \times C}$$

$$= \frac{1}{2 \times \pi \times 400 \times 10^3 \times 250 \times 10^{-12}}$$

$$= 1592 \; \Omega$$

$$\text{total impedance, } z = 30 + j\,377 - j\,1592$$
$$= 30 - j\,1215$$
$$= 1215\ \underline{/-88.6°}\ \Omega$$

$$\text{peak value of the fundamental current, } i_1 = \frac{100\ \text{V}}{1215\ \underline{/-88.6°}\ \Omega}$$
$$= 0.082\ \underline{/88.6°}\ \text{A}$$

$$\text{effective value of the fundamental current} = 0.082 \times 0.707$$
$$= 0.058\ \text{A}$$

Step 2. The Second Harmonic Component

$$\text{inductive reactance, } X_L = 2 \times 377 = 754\ \Omega$$

$$\text{capacitive reactance, } X_C = \frac{1592}{2} = 796\ \Omega$$

$$\text{total impedance, } z = 30 + j\,754 - j\,796\ \Omega$$
$$= 30 - j\,42$$
$$= 51.6\ \underline{/-54.5°}\ \Omega$$

peak value of the second harmonic current,
$$i_2 = \frac{-80\ \underline{/40°}\ \text{V}}{51.6\ \underline{/-54.5}\ \Omega} = -1.55\ \underline{/94.5°}$$
$$= 1.55\ \underline{/-85.5°}\ \text{A}$$

$$\text{effective second harmonic current, } I_2 = 0.707 \times 1.55 = 1.10\ \text{A}$$

Step 3. The Third Harmonic

$$\text{inductive reactance, } X_L = 3 \times 377 = 1131\ \Omega$$

$$\text{capacitive reactance, } X_C = \frac{1592}{3} = 531\ \Omega$$

$$\text{total impedance, } z = 30 + j\,1131 - j\,531$$
$$= 30 + j\,600$$
$$= 600.7\ \underline{/87.1°}\ \Omega$$

$$\text{peak value of the third harmonic current, } i_3 = \frac{40\ \underline{/-20°}\ \text{V}}{600.7\ \underline{/87.1°}\ \Omega}$$
$$= 0.067\ \underline{/-107.1°}\ \text{A}$$

$$\text{effective value of the third harmonic current, } I_3 = 0.707 \times 0.067$$
$$= 0.047\ \text{A}$$

$$\text{total nonsinusoidal current, } i = 0.082 \sin(\omega t + 88.6°)$$
$$+ 1.55 \sin(2\,\omega t - 85.50°) + 0.067 \sin(3\,\omega t - 107.1°)\ \text{A}$$

$$\text{effective nonsinusoidal current, } I = \sqrt{0.058^2 + 1.10^2 + 0.047^2}$$
$$= 1.1025\ \text{A}$$

total power dissipated $= I^2R$
$$= (1.1025 \text{ A})^2 \times 30 \ \Omega = 36.5 \text{ W}$$

The analysis of a parallel LCR circuit is similar to that of the series circuit. The individual branch currents for the fundamental and harmonic components are calculated first. By using the rules of complex algebra, the branch currents are combined to create the supply current that is drawn from the nonsinusoidal voltage source.

NONSINUSOIDAL CURRENT IN A SERIES-PARALLEL CIRCUIT

When a nonsinusoidal voltage source is applied across the series-parallel circuit of Fig. 19-7, we can determine the fundamental current components in each of the two branches. The fundamental line current is then the phasor sum of the branch currents. The procedure is repeated for each of the harmonic components. Finally the various currents are combined to obtain the effective value of the line current drawn from the source. These principles are illustrated in the following example:

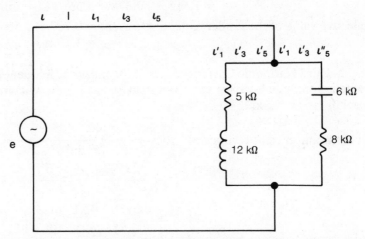

Fig. 19-7 Nonsinusoidal voltage applied to a series-parallel ac circuit.

In Fig. 19-7, the nonsinusoidal voltage source is expressed by the equation

$$e = 14.14 \sin (\omega t + 20°) + 3.535 \sin (3 \ \omega t + 40°)$$
$$- 1.414 \sin (5 \ \omega t - 60°) \text{ V}$$

where $\omega = 2 \ \pi f = 1000$ rad/s. The reactance values shown correspond to the fundamental frequency. Obtain the equation of the instantaneous line current and then calculate its effective value.

Step 1. Fundamental Current

$$\text{fundamental current component, } i_1' = \frac{14.14 \ \underline{/20°} \text{ V}}{5 + j \ 12 \text{ k}\Omega}$$

$$= \frac{14.14 \; \underline{/20°} \; (5 - j \, 12)}{5^2 + 12^2}$$

$$= \frac{14.14 \; \underline{/20°} \times 13 \; \underline{/-67.4°}}{169}$$

$$= 1.09 \; \underline{/-47.4°} \text{ mA}$$

fundamental current component, $i_1'' = \dfrac{14.14 \; \underline{/20°} \text{ V}}{8 - j \, 6 \text{ k}\Omega}$

$$= \frac{14.14 \; \underline{/20°} \times (8 + j \, 6)}{100}$$

$$= \frac{14.14 \; \underline{/20°} \times 10 \; \underline{/36.8°}}{100}$$

$$= 1.414 \; \underline{/56.8°} \text{ mA}$$

fundamental line current, $i_1 = 1.09 \; \underline{/-47.4°} + 1.414 \; \underline{/56.8°}$

$$= 0.738 - j \, 0.802 + 0.774 + j \, 1.28$$

$$= 1.512 + j \, 0.378 = 1.56 \; \underline{/14.04°} \text{ mA}$$

effective value of the fundamental line current $= 0.707 \times 1.56$
$$= 1.10 \text{ mA}$$

Step 2. Third Harmonic Current   At the third harmonic frequency, the new reactance of the inductor is $3 \times 12 = 36$ kΩ and the new reactance of the capacitor is $6/3 = 2$ kΩ.

third harmonic current component, $i_3' = \dfrac{3.535 \; \underline{/40°} \times (5 - j \, 36)}{5^2 + 36^2}$

$$= \frac{3.535 \; \underline{/40°} \times 36.3 \; \underline{/-82.1°}}{1321}$$

$$= 0.097 \; \underline{/-42.1°} \text{ mA}$$

third harmonic current component, $i_3'' = \dfrac{3.535 \; \underline{/40°} \times (8 + j \, 2)}{8^2 + 2^2}$

$$= \frac{3.535 \; \underline{/40°} \times 8.25 \; \underline{/14.0°}}{68}$$

$$= 0.43 \; \underline{/54°} \text{ mA}$$

third harmonic line current, $i_3 = 0.097 \; \underline{/-42.1°} + 0.43 \; \underline{/54°}$

$$= 0.072 - j \, 0.065 + 0.253 + j \, 0.348$$

$$= 0.325 + j \, 0.283$$

$$= 0.431 \; \underline{/41.05°} \text{ mA}$$

effective value of the third harmonic line current $= 0.707 \times 0.431$
$$= 0.305 \text{ mA}$$

Step 3. Fifth Harmonic Current At the fifth harmonic frequency the inductive reactance is $5 \times 12 = 60$ kΩ while the capacitive reactance is $6/5 = 1.2$ kΩ

$$\text{fifth harmonic current component } i_5' = \frac{-1.414 \; \underline{/-60°} \times (5 - \text{j } 60)}{5^2 + 60^2}$$

$$= \frac{-1.414 \; \underline{/-60°} \times 60.2 \; \underline{/-85.2°}}{3625}$$

$$= -0.0235 \; \underline{/-145.2°}$$

$$= 0.0235 \; \underline{/34.8°} \text{ mA}$$

$$\text{fifth harmonic current component, } i_5'' = \frac{-1.414 \; \underline{/-60°} \times (8 + \text{j } 1.2)}{8^2 + 1.2^2}$$

$$= \frac{-1.414 \; \underline{/-60°} \times 8.09 \; \underline{/8.53°}}{65.44}$$

$$= -0.175 \; \underline{/-51.47°}$$

$$= 0.175 \; \underline{/128.5°} \text{ mA}$$

$$\text{fifth harmonic line current, } i_5 = 0.0235 \; \underline{/34.8°} + 0.175 \; \underline{/128.5°}$$

$$= 0.0189 + \text{j } 0.0134 - 0.109 + \text{j } 0.137$$

$$= -0.0901 + \text{j } 0.1504$$

$$= 0.175 \; \underline{/120.1°} \text{ mA}$$

$$\text{effective value of the fifth harmonic line current} = 0.707 \times 0.175$$

$$= 0.124 \text{ mA}$$

Step 4 The expression for the instantaneous line current is

$$i = i_1 + i_3 + i_5$$
$$= 1.56 \sin (\omega t + 14.04°) + 0.431 \sin (\omega t + 41.05°)$$
$$+ 0.175 \sin (5 \; \omega t + 120.1°)$$

The effective line current is

$$I = \sqrt{1.10^2 + 0.305^2 + 0.124^2} = 1.15 \text{ mA}$$

CHAPTER SUMMARY

☐ **Fourier's Theorem**

Any finite, continuous, single-valued function, $f(t)$, that has a period of 2π radians (360°), can be expressed by the following series:

$$f(t) = a_0 + A_1 \sin (\omega t + \phi_1) + A_2 \sin (2 \; \omega t + \phi_2) + A_3 \sin (3 \; \omega t + \phi_3)$$
$$= a_0 + a_1 \cos \omega t + a_2 \cos 2 \; \omega t + a_3 \cos 3 \; \omega t + \ldots$$
$$+ b_1 \sin \omega t + b_2 \sin 2 \; \omega t + b_3 \sin 3 \; \omega t + \ldots$$

☐ Series Expressions for Various Nonsinusoidal Waveforms
Symmetrical square wave (Fig. 19-1D):

$$f(t) = \frac{4\,D}{\pi}\left(\sin\omega t + \frac{1}{3}\sin 3\,\omega t + \frac{1}{5}\sin 5\,\omega t + \frac{1}{7}\sin 7\,\omega t \ldots\right)$$

D = peak value of the symmetrical square wave

Negative sawtooth waveform (Fig. 19-1A):

$$f(t) = \frac{2\,D}{\pi}\left(\sin\omega t + \frac{1}{2}\sin 2\,\omega t + \frac{1}{3}\sin 3\,\omega t + \frac{1}{4}\sin 4\,\omega t + \ldots\right)$$

D = peak value of the negative sawtooth waveform

Modified sawtooth waveform (Fig. 19-5):

$$f(t) = \frac{D}{4} - \frac{2\,D}{\pi^2}\left(\cos\omega t + \frac{\cos 3\,\omega t}{3^2} + \frac{\cos 5\,\omega t}{5^2} + \frac{\cos 7\,\omega t}{7^2} + \ldots\right)$$

$$+ \frac{D}{\pi}\left(\sin\omega t - \frac{\sin 2\,\omega t}{2} + \frac{\sin 3\,\omega t}{3} \ldots\right)$$

D = peak value of the modified sawtooth waveform

Triangular waveform (Fig. 19-1C):

$$f(t) = \frac{8\,D}{\pi^2}\left(\cos\omega t + \frac{\cos 3\,\omega t}{3^2} + \frac{\cos 5\,\omega t}{5^2} + \frac{\cos 7\,\omega t}{7^2} + \ldots\right)$$

D = peak value of the triangular waveform

Half-wave rectification (Fig. 19-1F):

$$f(t) = \frac{2\,D}{\pi}\left(\frac{1}{2} + \frac{\pi}{4}\sin\omega t - \frac{1}{1\times 3}\cos 2\,\omega t - \frac{1}{3\times 5}\cos 4\,\omega t\right.$$

$$\left. - \frac{1}{5\times 7}\cos 6\,\omega t \ldots\right)$$

D = peak value of the half-rectified sine wave

Full-Wave Rectification (Fig. 19-1G):

$$f(t) = \frac{4\,D}{\pi}\left(\frac{1}{2} - \frac{1}{1\times 3}\cos 2\,\omega t - \frac{1}{3\times 5}\cos 4\,\omega t - \frac{1}{5\times 7}\cos 6\,\omega t \ldots\right)$$

D = peak value of the full-rectified sine wave

Effective values of the nonsinusoidal waveform:

$$\text{total effective current} = \sqrt{I_{DC}^2 + I_1^2 + I_2^2 + I_3^2 \ldots}$$

$$\text{total effective voltage} = \sqrt{V_{DC}^2 + V_1^2 + V_2^2 + V_3^2 \ldots}$$

□ Application of a Nonsinusoidal Voltage Source to Ac Circuits

The individual currents must be calculated separately for each of the fundamental and harmonic components contained in the waveform of the voltage source. These currents then are combined to obtain the equation of the instantaneous current at any point in the circuit. From a knowledge of the instantaneous equations, the effective values of the currents can be calculated.

20
Electrical Measurements

WHEN STUDYING THE different types of meter, you will, in particular, learn:

☐ About the construction of a moving-coil meter movement and how it employs the motor effect to achieve its deflection.

☐ How the moving-coil instrument is adapted to behave as an ammeter for measuring the current.

☐ How the moving-coil instrument is adapted to behave as a voltmeter to measure potential differences.

☐ The effect that the resistance of a voltmeter has on the voltage values being recorded.

☐ How the moving-coil instrument is adapted to behave as an ohm-meter to measure resistance.

☐ How to use Wheatstone Bridge to obtain accurate measurements of resistance.

☐ About the principles of moving-iron ammeters and voltmeters as well as their advantages and disadvantages with respect to moving-coil instruments.

☐ The use of the thermocouple instrument to measure a high-frequency current.

☐ How an electrodynamometer meter movement is used to measure power.

The most common type of indicating instrument is fitted with a pointer that shows on a scale the value the quantity being measured. Such an instrument has a moving system that is carried by a hardened steel spindle with its ends tapered and highly polished. The tapered ends form pivots that rest in hollow-ground sapphire bearings set in steel screws.

INTRODUCTION TO INDICATING INSTRUMENTS

Indicating instruments of the above type all possess three essential features:

1. A deflecting device that enables a mechanical force to be exerted by the voltage, current, or power being measured.
2. A controlling device that ensures that the amount of the deflection is dependent on the magnitude of the measured quantity.
3. A damping device that prevents oscillation of the moving system and enables the final position to be reached quickly.

The various devices used depend on the particular type of instrument and its purpose.

MOVING-COIL (D'ARSONVAL) METER MOVEMENT

Referring to Fig. 20-1, the deflecting device of the moving-coil meter movement consists of a rectangular coil. C, that is made of thin insulated copper wire wound on a light aluminum former. This frame is carried by a spindle that pivots in jewelled bearings. Current is led into and out of the coil by the spiral hairsprings, H, which are the controlling device and therefore provide the restoring torque. The coil is free to move in the gaps between the permanent magnet pole-pieces, P, and a soft iron cylinder, A, that is carried normally by a non-magnetic bridge attached to P. The purpose of the soft iron cylinder is

1. To concentrate the magnetic flux by reducing the amount of the air-gap.
2. To produce a radial magnetic field with a uniform flux density. This also is achieved by special shaping of the pole pieces, P.

Let the coil consist of N turns, each with a length of L meters and a width of d meters.

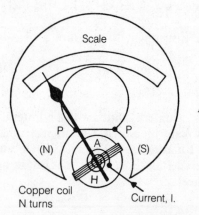

Fig. 20-1 Moving-coil meter movement.

In a radial field the torque, T, exerted on the coil is

$$T = BINLd = BANI \text{ newton-meters}$$

where

I = current flowing through coil in amperes
B = flux density in teslas
A = area of coil in square meters

The restoring torque, T′, provided by the hairsprings, is directly proportional to the amount of the deflection, $\theta°$. Therefore $T' = k\theta$ where k is a constant, and in the final equilibrium position,

$$T = T'$$
$$BANI = k\theta$$
$$I = \frac{k\theta}{BAN} = K\theta$$

where K is the constant of proportionality for the entire meter movement. The deflection is directly proportional to the current so that the scale is linear with the divisions evenly spaced.

As the coil swings towards the equilibrium position, it gains momentum that could create unwanted oscillation. The required damping is introduced by winding the coil on the aluminum frame. As the coil rotates, eddy currents are induced into the frame, and these absorb the kinetic energy of the system. When the coil reaches its equilibrium position, it has no energy to swing further, and because the deflecting and restoring torques are balanced, efficient damping is obtained.

The sensitivity of the moving coil meter movement is determined by the amount of current required to provide full-scale deflection (the lower the value of current, the greater the movement's sensitivity). High sensitivity is achieved by using a strong permanent magnet to produce a large flux density; in addition, many turns of fine copper wire are wound on a light aluminum frame and then attached to delicate hairsprings. The full-scale deflection current for such a meter movement is 20 μA or less. By using fewer turns of the larger wire size together with stronger hairsprings, the full-scale deflection current can be increased to a few mA. The limitation on the current then becomes the physical size of the meter movement.

Because the sensitivity and the full-scale deflection current are related inversely, the sensitivity is measured directly in terms of the current's reciprocal. With current equivalent to volts divided by ohms, the units of sensitivity will be ohms per volt. For example, if the meter movement's full-scale deflection current is 50 μA, the sensitivity is

$$1/50 \ \mu A = \frac{1}{50 \times 10^{-6}} = 20,000 \text{ ohms per volt}$$

Therefore

$$\text{sensitivity (ohms per volt)} = \frac{1}{\text{full-scale deflection current (amperes)}}$$

and

$$\text{full-scale deflection current (amperes)} = \frac{1}{\text{sensitivity (ohms per volt)}}$$

The moving coil meter movement is therefore basically a microammeter or milliammeter. Typically the resistance of the coil is on the orders of tens of ohms, and therefore the meter movement can be calibrated to measure only millivolts.

The passage of current through the coil raises its temperature and alters its resistance. It is normal practice to add a swamp resistor in series with the coil. The swamp resistor has a negative temperature coefficient that counteracts the positive temperature coefficient of the copper coil. It is also used to bring the total resistance (coil resistance plus swamp resistance) to some suitable value such as 50 Ω. It must be remembered that in order to measure current at a particular point, the circuit must be broken and the meter inserted in the break. The total 50 Ω resistance of the meter movement will be in series with the remainder of the circuit and therefore must affect the reading to a certain extent. The amount of the error depends on the value of the circuit's total resistance in comparison with the 50 Ω resistance of the meter movement.

The main advantages of the moving coil meter movement are:

1. Uniform scale
2. High sensitivity
3. High degree of shielding from stray magnetic fields

The principal disadvantages are:

1. Only suitable for dc measurement
2. Relatively expensive when compared with the moving-iron instrument

Example 20-1

The sensitivity of a moving coil meter movement is 1000 Ω/V. What is the value of the full-scale deflection current?

Solution

$$\text{full-scale deflection current} = \frac{1}{1000 \ \Omega/\text{V}}$$

$$= \frac{1}{1000} \ \text{A} = 1 \ \text{mA}$$

THE AMMETER

In order to convert the basic moving-coil meter movement into an instrument capable of measuring hundreds of milliamperes or even amperes, a shunt resistor is connected across the series combination of the meter movement and the swamp resistor, (Fig. 20-2). Because the shunt resistor has a low value, most of the current to be measured is diverted through R_{sh}, and only a small fraction passes through the meter movement to provide the required deflection. Then

$$I = I_m + I_{sh}$$
$$I_{sh} \times R_{sh} = I_m \times R_m;$$

and

$$R_{sh} = \frac{I_m \times R_m}{I_{sh}} = \frac{I_m \times R_m}{I - I_m}$$

where

I = current to be measured
I_m = current flowing through the meter movement
R_{sh} = shunt resistance
R_m = total meter movement resistance

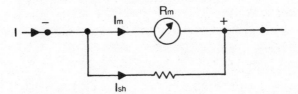

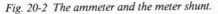

Fig. 20-2 The ammeter and the meter shunt.

Figure 20-3 shows an instrument that has a number of shunts for different current ranges. However, there is a danger that when switching from range to range, the meter movement will be placed directly in the circuit without the protection of a shunt. This can be avoided by making the switch of a make-before-break type. Alternatively, the instrument can use a Ayrton shunt, (Fig. 20-4), which ensures that the meter cannot be placed in the circuit without the presence of a shunt.

Because of its very low resistance, the ammeter must never be placed directly across a voltage source. When measuring an unknown current, the meter should be switched initially to the highest range. If the reading is then too small, the meter can be switched to a lower range; in this way the meter movement is protected from excessive current.

Example 20-2

In Fig. 20-3, the full-scale deflection current of the meter movement is 1 mA, and its resistance is 50 Ω. Calculate the values of the shunt resistors for the ranges of (a) 0 – 10 mA (b) 0 – 100 mA and (c) 0 – 1 A.

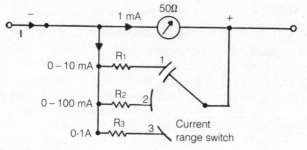

Fig. 20-3 Multirange ammeter.

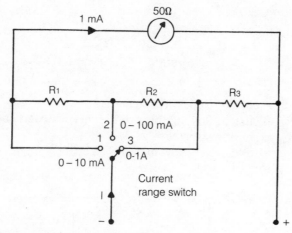

Fig. 20-4 Multirange ammeter with Ayrton shunt.

Solution

(a) The voltage across the meter movement = 1 mA × 50 Ω = 50 mV. For the full scale current of 10 mA, the shunt current is 10 mA − 1 mA = 9 mA. The value of the shunt resistor is

$$R_{sh} = \frac{50 \text{ mV}}{9 \text{ mA}} = 5.5\dot{5} \text{ } \Omega$$

The total resistance of the instrument is now

$$\frac{50 \times 5.5\dot{5}}{50 + 5.5\dot{5}} = 5 \text{ } \Omega$$

When the current to be measured is 5 mA, the voltage across the instrument is 5 mA × 5 Ω = 25 mV, and the current through the meter movement is 25 mV/50 Ω = 0.5 mA. The pointer therefore correctly indicates half of the full-scale deflection.

(b) $1 = 100$ mA, $I_m = 1$ mA, $R_m = 50 \, \Omega$

$$R_{sh} = \frac{1 \text{ mA} \times 50 \, \Omega}{100 \text{ mA} - 1 \text{ mA}} = 0.505 \, \Omega$$

(c) $I = 1$ A $= 1000$ mA

$$R_{sh} = \frac{1 \text{ mA} \times 50 \, \Omega}{1000 \text{ mA} - 1 \text{ mA}} = 0.05005 \, \Omega$$

These shunts normally are manufactured from resistance wire with a low temperature coefficient. For currents greater than 20 A, the shunts are too large to be placed within the instrument and therefore are connected externally.

Example 20-3

In Fig. 20-4, an Ayrton shunt is used for the same meter movement and current ranges as in Example 20-2. Calculate the required values for R_1, R_2, and R_3.

Solution

The problem is best solved by using the current division rule.
For the 0 – 10 mA range,

$$\frac{R_1 + R_2 + R_3}{50} = \frac{1 \text{ mA}}{9 \text{ mA}}$$

$$R_1 + R_2 + R_3 = \frac{50}{9}$$

$$R_2 + R_3 = \frac{50}{9} - R_1$$

For the 0 – 100 mA range,

$$\frac{R_2 + R_3}{R_1 + 50} = \frac{1 \text{ mA}}{99 \text{ mA}}$$

$$R_2 + R_3 = \frac{R_1 + 50}{99}$$

Therefore,

$$\frac{50}{9} - R_1 = \frac{R_1 + 50}{99}$$

$$550 - 99R_1 = R_1 + 50$$
$$R_1 = 5 \, \Omega$$

and

$$R_2 + R_3 = \frac{50}{9} - 5 = \frac{5}{9} = 0.55 \, \Omega$$

For the 0 – 1 A range,

$$\frac{R_3}{R_1 + R_2 + 50} = \frac{1}{999}$$

$$999R_3 = R_2 + 55$$
$$999R_3 - R_2 = 55$$

Therefore,

$$1000\ R_3 = 55\frac{5}{9} = 55.5\dot{5}$$

$$R_3 = 0.05\dot{5}\ \Omega$$

Then

$$R_2 = 0.5\dot{5} - 0.05\dot{5} = 0.5\ \Omega$$

The required values are therefore $R_1 = 5\ \Omega$, $R_2 = 0.5\ \Omega$, $R_3 = 0.5\dot{5}\ \Omega$.

THE VOLTMETER

As already indicated, the moving-coil meter movement is basically a millivoltmeter or microvoltmeter. In order to adapt the movement for higher voltage ranges, it is necessary to connect a series multiplier resistor. Most of the voltage to be measured is then dropped across this high value series resistor, and only a small part appears across the meter movement to provide the necessary deflection.

In the example of Fig. 20-5, the meter movement once again has a total resistance of 50 Ω and requires a full-scale deflection current of 1 mA. In order to adapt the instrument for the 0 — 10 V range, the total resistance of the voltmeter must be

$$\frac{10\ V}{1\ mA} = 10\ k\Omega$$

and the value of the multiplier resistor is 10 kΩ — 50 Ω = 9950 Ω. In equation form,

$$R_s = \frac{V}{I_m} - R_m$$
$$= V \times (\text{sensitivity in } \Omega/V) - R_m$$

where

R_s = value of the series multiplier resistor
V = voltage indicated by full-scale deflection
I_m = full-scale deflection current
R_m = resistance of meter movement

In a multirange voltmeter it is possible to use a separate multiplier resistor for each range. However, it is simpler to use a series of resistors as shown in Fig. 20-6.

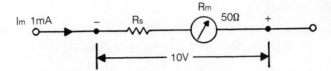

Fig. 20-5 *Principle of the voltmeter.*

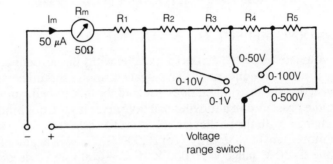

Fig. 20-6 *The multirange voltmeter.*

The moving-coil meter movement does not respond to ac directly. For ac measurements, the alternating voltage usually is rectified by a bridge circuit so that the resulting dc can deflect the moving-coil meter movement. The instrument then can be calibrated to measure the effective value of the ac voltage.

Example 20-4

In Fig. 20-6, the meter movement has a resistance of 50 Ω and a sensitivity of 20,000 Ω/V. The voltage ranges are 0–1 V, 0–10 V, 0–50 V, 0–100 V, 0–500 V. Calculate the required values for the resistors R_1, R_2, R_3, R_4, and R_5.

Solution

For the 0–1 V range

$$R_1 = (20000 \ \Omega/V \times 1 \ V) - 50 \ \Omega = 19950 \ \Omega$$

For the 0 – 10 V range

$$R_1 + R_2 = (20,000 \times 10) - 50 = 199,950 \ \Omega$$
$$R_2 = 199,950 - 19,950 = 180,000 \ \Omega$$

For the 0–50 V range,

$$R_1 + R_2 + R_3 = (20,000 \times 50) - 50 = 999,950 \ \Omega$$
$$R_3 = 999,950 - 199,950 = 800,000 \ \Omega$$

0 – 100 V Range

$$R_1 + R_2 + R_3 + R_4 = (20{,}000 \times 100) - 50 = 1{,}999{,}950 \ \Omega$$
$$R_4 = 1{,}999{,}950 - 999{,}950 = 1{,}000{,}000 \ \Omega$$

0 – 500 V Range

$$R_1 + R_2 + R_3 + R_4 + R_5 = 20{,}000 \times 500 - 50 = 9{,}999{,}950 \ \Omega$$
$$R_5 = 9{,}999{,}950 - 1{,}999{,}500 = 8{,}000{,}000 \ \Omega.$$

THE LOADING EFFECT OF A VOLTMETER

The voltmeter is placed across (in parallel with) the component whose dc voltage drop is to be measured. The ideal voltmeter should have infinite resistance so that the instrument does not load the circuit being monitored. However, the resistance of a moving-coil voltmeter is far from infinite, and its value changes with the particular voltage range selected. For example, if the meter movement has a sensitivity of 20,000 Ω/V, the instrument's resistance on the 0 – 10 V range is $20{,}000 \times 10 \ \Omega = 200$ kΩ, while on the 0 – 100 V range, the resistance increases to $20{,}000 \times 100 \ \Omega = 2$ MΩ. The worst loading effect by such a voltmeter therefore occurs in high resistance, low-voltage circuits, as shown in Fig. 20-7A.

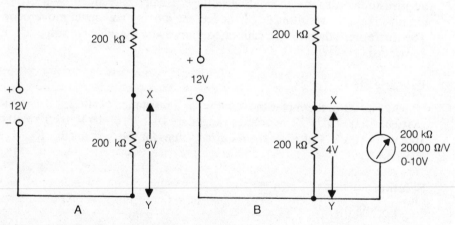

Fig. 20-7 *Loading effect of a voltmeter.*

The instrument should measure a voltage drop of 6 V between the points X and Y; the voltmeter therefore is switched to the 0 – 10 V range and then connected across the lower 200 kΩ resistor, (Fig. 20-7B). However, the voltmeter's resistance is also 200 kΩ so that the total resistance of the voltmeter and the resistor in parallel is only 200 kΩ/2 = 100 kΩ. Using the voltage-division rule, the reading of the voltmeter is

$$12 \text{ V} \times \frac{100 \text{ k}\Omega}{200 \text{ k}\Omega + 100 \text{ k}\Omega} = 4 \text{ V}$$

The voltmeter loading effect therefore has reduced the voltage between X and Y from 6 V and 4 V.

The field-effect transistor (FET) voltmeter has a high resistance (several megohms) that is independent of the voltage range chosen. This voltmeter has a minimum loading effect.

Example 20-5

In Fig. 20-7, the voltage drop between X and Y is measured by a voltmeter whose sensitivity is 1,000 Ω/V. What is the reading of the voltmeter when switched to the 0–10 V range?

Solution

$$\text{resistance of voltmeter} = 1000\ \Omega/\text{V} \times 10\ \text{V} = 10000\ \Omega$$
$$= 10\ \text{k}\Omega$$

total resistance of the voltmeter and the lower

$$200\ \text{k}\Omega\ \text{resistor in parallel} = \frac{10 \times 200}{10 + 200} = 9.524\ \text{k}\Omega$$

Using the voltage-division rule, the voltmeter reading is

$$12\ \text{V} \times \frac{9.524\ \text{k}\Omega}{200\ \text{k}\Omega + 9.524\ \text{k}\Omega} = 0.545\ \text{V}$$

THE OHMMETER

The moving-coil meter movement can be adapted for resistance measurements by including additional resistors and one or more primary cells. Because these cells alone must provide the necessary deflection, it is essential that resistance movements are only made with all power removed from the circuit.

Figure 20-8A shows the circuit of a practical ohmmeter that uses a meter movement with a sensitivity of 20,000 Ω/V and a resistance of 2,000 Ω. The full-scale deflection current therefore will be 50 μA, which must flow through the meter movement when the red and black probes are shorted together so that the measured resistance is zero ohms. Shorting the probes together is the initial step in the use of the ohmmeter; the value of the zero-adjust resistor, R_6, is set to produce the full-scale deflection current that will coincide with the reading of zero ohms on the right-hand side of the resistance scale (Fig. 20-8B).

With the range switch in the R $\times$ 1 position the value of R_6 is

$$\frac{1.5\ \text{V}}{50\ \mu\text{A}} - 1138\ \Omega - 21850\ \Omega - 2000\ \Omega$$

$$= 30\ \text{k}\Omega - 24988\ \Omega$$
$$= 5012\ \Omega$$

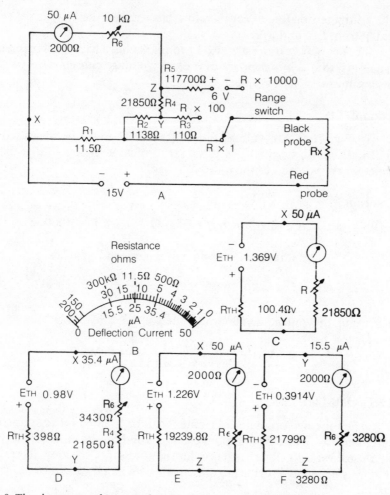

Fig. 20-8 The ohmmeter and its equivalent circuits for examples 20-6 and 20-7.

The value of R_1 is significant because 11.5 Ω is the value in the center of the scale so that when the unknown resistance, R_x, is 11.5 Ω, the current through the meter movement must be 50 μA/2 = 25 μA. On the left side of the scale, the deflection is zero and therefore R_x is an open circuit of infinite ohms. This means that the resistance scale is nonlinear as shown in Fig. 20-8B. The R × 1 range is used to measure resistances between zero and about 200 Ω, although it is difficult to measure accurately on the left-hand side of the scale.

When R_x is 11.5 Ω, the cell voltage of 1.5 V divides equally between R_x and R_1 (ignoring the shunting effect of $R_M + R_2 + R_4 + R_6 = 30$ kΩ across R_1). The voltage across R_1 is therefore 1.5 V/2 = 0.75 V, which will drive the necessary current of 0.75 V/30 kΩ = 25 μA through the meter movement.

In the R × 100 range position, the value of R_6 must be reset; R_3 is included then so that when R_x is $11.5 \times 100 = 1,150$ Ω, the current through the meter movement is 25 μA. This range is used to measure values of R_x between 200 Ω and 20,000 Ω.

On the R × 10,000 range (R_x between 20,000 Ω and 2 MΩ), the additional 6 V cell must be added in order to drive the required current through the meter movement. The value of R_5 is such that when R_x is $11.5 \times 10,000 = 115,000$ Ω, the deflection current is again 25 μA.

Example 20-6

In Fig. 20-8A, calculate the value for R_6 on the R × 100 range. If R_x is 500 Ω, what is the value of the deflection current through the meter movement?

Solution

With the probes shorted, Thévenize the circuit between the points X and Y (Fig. 14-8C). They by the voltage division rule

$$E_{TH} = 1.5 \text{ V} \times \frac{(11.5 \text{ Ω} + 1138 \text{ Ω})}{11.5 \text{ Ω} + 1138 \text{ Ω} + 110 \text{ Ω})}$$

$$= \frac{1.5 \times 1149.5}{1259.5} = 1.369 \text{ V.}$$

$$R_{TH} = \frac{110 \times 1149.5}{110 + 1149.5} = 100.4 \text{ Ω.}$$

Then

$$R_6 = \frac{1.369 \text{ V}}{50 \text{ μA}} - 21850 \text{ Ω} - 2000 \text{ Ω} - 100.4 \text{ Ω}$$

$$= 3430 \text{ Ω}$$

When the resistance of 500 Ω is connected between the probes (Fig. 14-8D),

$$E_{TH} = 1.5 \text{ V} \times \frac{(11.5 \text{ Ω} + 1138 \text{ Ω})}{11.5 \text{ Ω} + 1138 \text{ Ω} + 110 \text{ Ω} + 500 \text{ Ω}}$$

$$= 1.5 \times \frac{1149.5}{1759.5} = 0.98 \text{ V}$$

and

$$R_{TH} = \frac{610 \times 1149.5}{610 + 1149.5} = 398 \text{ Ω}$$

The deflection current is

$$\frac{0.98 \text{ V}}{398 \text{ Ω} + 3430 \text{ Ω} + 2000 \text{ Ω} + 21,850 \text{ Ω}} = 35.4 \text{ μA}$$

Example 20-7

In Fig. 14-8A, calculate the value required for R_6 on the R × 10,000 range. If R_x is 300 kΩ, what is the value of the deflection current through the meter movement?

Solution

With the probes shorted, Thévenize the circuit between the points X and Z (Fig. 14-8E). By the voltage division rule,

$$E_{TH} = 7.5 \text{ V} \times \frac{(11.5 \text{ } \Omega + 1138 \text{ } \Omega + 21{,}850 \text{ } \Omega)}{11.5 \text{ } \Omega + 1{,}138 \text{ } \Omega + 21{,}850 \text{ } \Omega + 117{,}700 \text{ } \Omega}$$

$$= 7.5 \text{ V} \times \frac{22{,}999.5 \text{ } \Omega}{140{,}699.5 \text{ } \Omega} = 1.226 \text{ V}$$

$$R_{TH} = \frac{22{,}999.5 \times 117{,}700}{140{,}699.5} = 19{,}239.8 \text{ } \Omega$$

Then

$$R_6 = \frac{1.226 \text{ V}}{50 \text{ } \mu A} - 19{,}239.8 \text{ } \Omega - 2{,}000 \text{ } \Omega = 3{,}280 \text{ } \Omega$$

When the resistance of 300 kΩ is connected between the probes (Fig. 14-8F),

$$E_{TH} = 7.5 \text{ V} \times \frac{11{,}999.5 \text{ } \Omega}{440{,}699.5 \text{ } \Omega} = 0.3914 \text{ V}$$

$$R_{TH} = \frac{22.999.5 \times 417{,}700}{440{,}699.5} = 21{,}799 \text{ } \Omega$$

The deflection current is

$$\frac{0.3914 \text{ V}}{21{,}799 \text{ } \Omega + 3{,}280 \text{ } \Omega + 2{,}000 \text{ } \Omega} = 15.5 \text{ } \mu A$$

THE WHEATSTONE BRIDGE

This is a precision method of measuring resistances over a wide range of values. The circuit is shown in Fig. 20-9 and consists of four resistance arms, a sensitive current meter (G), and a dc voltage source. R_3 is an accurately calibrated variable resistor while each of the fixed resistors R_1, R_2, has possible values of only 1 Ω, 10 Ω, 100 Ω, or 1,000 Ω. The ratio of $R_1 : R_2$ then can be set for 0.001, 0.01, 0.1, 1, 10, 100, or 1000.

To measure the unknown resistance, R_x, it is necessary to *balance* the bridge; this is accomplished by selecting the required ratio for $R_1 : R_2$ and then adjusting R_3 until the reading of G is zero.

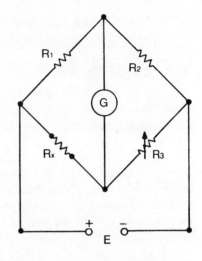

Fig. 20-9 The Wheatstone bridge.

When the bridge is balanced, the voltage drops across R_2 and R_3 must be equal. Therefore by the voltage division rule,

$$\frac{E \times R_2}{R_1 + R_2} = \frac{E \times R_3}{R_x + R_3}$$

$$R_x = R_3 \times \frac{R_1}{R_2}$$

The measured value of R_x can therefore range from the lowest value of R_3 divided by 1000 to 1000 times the highest values of R_3.

Example 20-8

In Fig. 14-9, $R_1 = 1{,}000 \ \Omega$ and $R_2 = 10 \ \Omega$. When the bridge is balanced, R_3 is 8.2 Ω. What is the value of the unknown resistor, R_x? If the values of R_1 and R_2 are interchanged, what is the new value of R_x?

Solution

$$\text{unknown resistance, } R_x = 8.2 \times \frac{1000}{10} = 8{,}200 \ \Omega = 8.2 \ k\Omega$$

When the resistors, R_1, R_2 are interchanged, the new value of

$$R_x = 8.2 \times \frac{10}{1000} = 0.082 \ \Omega$$

MOVING-IRON AMMETERS AND VOLTMETERS

These instruments can be divided into two types:

1. The attraction type, in which a piece of soft iron is attracted towards a solenoid.

2. The repulsion type, in which two parallel, soft iron vanes are magnetized inside a solenoid and therefore repel each other.

Attraction Moving-Iron Meter

This instrument is illustrated in Fig. 20-10A. The current to be measured flows through the coil and sets up a magnetic flux. This field magnetizes the soft iron disc and draws it into the center of the coil. The force acting on the iron depends on the flux density and on the coil's magnetic field intensity, both of which are proportional to the current. The torque and the deflection therefore are not proportional to the current alone (as in the moving-coil meter movement) but to the square of the current. The result is a nonlinear scale that is cramped at the low end but open at the high end. However, by careful shaping of the iron disc, the scale's linearity can be improved.

The restoring torque of the controlling device is supplied by a spring or sometimes by gravity. The damping device is commonly obtained by the use of an air piston.

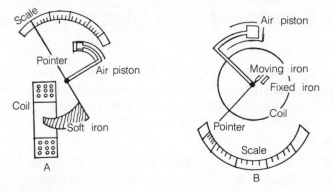

Fig. 20-10 Moving-iron instruments.

Repulsion Moving-Iron Meter

Two iron vanes are situated axially in a short solenoid as shown in Fig. 20-10B. One vane is fixed while the other is movable and attached to a pivot that also carries the pointer. When the current flows through the coil, the vanes are equally magnetized and repel each other. The repulsion creates a deflection of the needle on the scale, which is calibrated for direct reading. The force of repulsion is dependent on the flux density of each iron vane and is directly proportional to the square of the current; therefore the scale is again nonlinear. The restoring and damping systems are similar to those of the attraction type.

In the measurement of ac, the deflecting torque for both the attraction and repulsion meters is proportional to the square of the instantaneous current and therefore varies from zero to a peak value at a rate equal to twice the alternating frequency. Because of its inertia, the moving system takes up a position corresponding to the mean torque. The deflection is then proportional to the square of the current's rms value.

Unlike the moving-coil instrument, the moving iron meter can be used for both dc and ac measurements. If the current contains both dc and ac components, the reading is equal to

$$\sqrt{I_{dc}^2 + I_{rms}^2}$$

Because the strength of the deflection torque depends on the number of ampere turns for the coil, it is possible to arrange for various ranges by winding different numbers of turns on the coil. Also, by varying the type of wire used, the meter's resistance can be changed so that there is no need for shunt and swamp resistances.

The moving-iron milliammeter can be converted to a voltmeter by adding a suitable noninductive series resistor.

Apart from their advantage of being able to measure both dc and ac, the moving-iron instruments also are relatively cheap and robust, but their operation is only satisfactory at line frequencies; moreover, they have low sensitivity, are affected by stray magnetic fields, and are liable to hysteresis error when used to measure ac.

Example 20-9

A moving-iron meter needs 450 ampere turns to provide full-scale deflection. How many turns on the coil are required for the 0 – 10 A range? For a voltage scale of 0 – 300 V with a current of 15 mA, how many turns are required and what is the total resistance of the instrument?

Solution

$$\text{number of turns for the 0 – 10 A range} = \frac{450}{10} = 50 \text{ turns}$$

$$\text{number of turns for the 0 – 300 V range} = \frac{450}{0.015}$$

$$= 30{,}000 \text{ turns}$$

$$\text{total resistance} = \frac{300 \text{ V}}{15 \text{ mA}} = 20 \text{ k}\Omega$$

THE THERMOCOUPLE METER

The principle of this type of meter is based on the Seebeck effect, which was discovered originally in 1821. If a circuit consisting of different metals is at the same temperature throughout, there is no resulting EMF. However, if a junction between two dissimilar metals is maintained at a different temperature from the rest of the circuit, a *thermoelectric EMF* is generated, and a current then can be driven through a conventional moving-coil meter movement.

The construction of the meter is shown in Fig. 20-11. The current to be measured passes between X and Y, raising the temperature of the beater wire. Attached to the center of this wire is the thermocouple junction, J; bismuth and antimony commonly are used as the dissimilar metals although many other combinations are possible. As the temperature rises, the thermo-electric EMF increases and drives a greater current through the meter movement. The amount of deflection on the scale depends on a heating effect that is proportional to the square of the measured current. The meter scale is therefore nonlinear so that it is cramped at the low end and open at the high end.

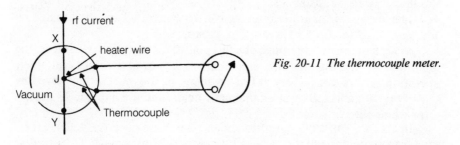

Fig. 20-11 The thermocouple meter.

This type of "current-squared" meter is suitable for reading both dc and the rms value of ac. Its particular importance lies in the measurement of radio-frequency currents such as occur in the antenna systems of transmitters (Fig. 20-12). Once this type of meter has been calibrated, the calibration is accurate from dc up to microwave frequencies.

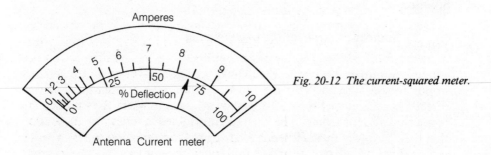

Fig. 20-12 The current-squared meter.

Example 20-10

An rf current of 10 A(rms) provides full-scale deflection of a thermocouple ammeter. What percentage of full-scale deflection corresponds to a current of 4 A(rms)?

Solution

The amount of deflection is proportional to the square of the current.

Therefore the deflection corresponding to 4 A is

$$100 \times \left(\frac{4\ A}{10\ A}\right)^2 = 16\% \text{ of the deflection}$$

Example 20-11

An rf current of 5 A(rms) corresponds to the full-scale deflection on a thermocouple ammeter. What value of current will produce 60% of the full-scale deflection?

Solution

The current is proportional to the square root of the deflection. Therefore the current corresponding to 60% deflection is

$$5\ A \times \sqrt{\frac{60}{100}} = 3.87\ A$$

ELECTRODYNAMOMETER MOVEMENT — THE WATTMETER

In the moving-coil meter movement, the flux associated with the permanent magnet is fixed in direction; consequently this type of meter (unless used in conjunction with a rectifier) can be used only for dc measurements. However, if the permanent magnet is replaced by an electromagnet, the direction of the flux can be reversed. This is shown in the *electrodynamometer* movement of Fig. 20-13A, in which two fixed coils F and F', provide the electromagnet. The moving coil, M, is then carried by a spindle, and the controlling torque is exerted by spiral hairsprings H and H', which also serve to lead the current into and out of the moving coil.

For dc measurements the electrodynamometer movement has no advantage over the D'Arsonval type. Compared with ac moving-iron instruments, dynamometer ammeters and voltmeters are less sensitive and more expensive, so they are rarely used. However, dynamometer wattmeters are important because they are the most common way of measuring power directly in ac circuits. Figure 20-13B shows the way in which the wattmeter is connected into the circuit. The fixed coils are joined in series with the load so that they carry the instantaneous load current. The moving coil is in series with a high value noninductive multiplier resistor, so that the current through the moving coil is proportional to and nearly in phase with the source voltage. The instantaneous torque on the moving coil is proportional to the product of the instantaneous current through the fixed (current) coils and the instantaneous current through the moving (voltage) coil and is therefore proportional to the instantaneous power taken by the load.

The controlling torque provided by the hairsprings depends directly on the deflection, which is therefore proportional to the power; consequently, electrodynamometer wattmeters have linear scales.

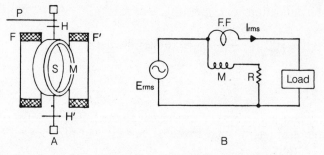

Fig. 20-13 The electro-dynamometer movement and the wattmeter.

If the load contains both resistance and reactance, the instantaneous power waveform is a second harmonic curve as shown in Fig. 10-1. When the curve is above the zero horizontal line, the torque on the moving coil is clockwise, but its direction is reversed when the coil falls below the line. The inertia of the voltage coil and the pointer does not allow the deflection to follow the variations in the instantaneous torque. The rest position of the pointer, P, therefore will represent the average (true) power taken by the load; in other words, the wattmeter automatically takes into account the power factor of the circuit. Therefore true power recorded by the wattmeter is

$$E \times I \times \text{power factor} = E \times I \times \cos \phi$$

where

E = rms value of the source voltage
I = rms value of the load current
ϕ = phase angle between the source voltage and the load

current

When the load is entirely reactive, the clockwise and counterclockwise torques exactly balance so that the wattmeter reads zero. However, heavy currents can be flowing in the fixed and moving coils so that the instrument must have both voltage and current ratings.

The power factor of the load can be determined by dividing the wattmeter reading by the product of the source voltage and the load current. However, power-factor meters exist to indicate continuously the value of the power factor and whether it is leading or lagging.

Example 20-12

Readings as taken at an ac source are voltmeter 110 V, ammeter 6.2 A, and wattmeter 590 W. What are the values of the power factor and the phase angle between the source voltage and the inductive load current?

Solution

$$\text{power factor} = \frac{\text{true power}}{\text{apparent power}}$$

$$= \frac{590 \text{ W}}{110 \text{ V} \times 6.2 \text{ A}} = 0.865, \text{ lagging}$$

$$\text{phase angle, } \phi = \text{inv cos } 0.865 = +30.1°$$

CHAPTER SUMMARY

☐ Moving-coil Meter Movement

$$\text{torque exerted on coil} = \text{BINLd} = \text{BANI meter-newtons}$$
$$\text{current, I} = \text{K} \times \theta° \text{ (deflection)}$$

$$\text{sensitivity (ohms per volt)} = \frac{1}{\text{full-scale deflection current (amperes)}}$$

☐ Ammeter

$$I = I_m + I_{sh}$$
$$I_{sh} \times R_{sh} = I_m \times R_m$$
$$R_{sh} = \frac{I_m \times R_m}{I_{sh}} = \frac{I_m \times R_m}{I - I_m}$$

☐ Voltmeter

$$R_s = \frac{V}{I_m} - R_m$$

☐ Wheatstone Bridge

$$R_x \times R_2 = R_3 \times R_1$$
$$R_x = R_3 \times \frac{R_1}{R_2}$$

☐ Current-squared (Thermocouple) Meter

$$\text{meter deflection } \alpha \text{ (current)}^2$$
$$\text{current } \alpha \sqrt{\text{meter deflection}}$$

☐ Wattmeter

$$\text{true power in watts} = E \times I \times \text{power factor}$$

$$\text{power factor, p.f.} = \frac{\text{wattmeter reading}}{E \times I}$$

Index